Schools Council
Research Studies

Science 5–13:
a Formative Evaluation

The Science 5–13 Project, sponsored by the
Schools Council, the Nuffield Foundation,
the Scottish Education Department and by
the Plastics Institute, was based at the
Bristol University Institute of Education
from 1967 to 1974. The author of this
report on the development of the curriculum
materials and their formative evaluation
was Project Evaluator from 1967 to 1973.

Schools Council
Research Studies

Science 5–13: a Formative Evaluation

Wynne Harlen

M

Macmillan Education

First published 1975

SBN 333 17797 5

Published by MACMILLAN EDUCATION LTD
London and Basingstoke

Associated companies and representatives
throughout the world

Printed in Great Britain by
Hazell Watson & Viney Ltd
Aylesbury, Bucks

Contents

Tables and figures

Tables

Figures

Acknowledgements

Science 5–13 was privileged to have the co-operation and assistance of a great many people in producing and evaluating its materials. The work described in this report could not have been carried out without the hard work of the teachers and others who took part. Together with the rest of the project team I should like to express grateful thanks for the help so willingly given to Science 5–13 by teachers, advisers, inspectors and administrators. I am particularly indebted to the trial area representatives for their work in organizing the evaluation trials and visiting trial classes.

I should also like to thank the members of the project's consultative committee for their guidance and support with regard to the evaluation. I am extremely grateful to the members of the Science 5–13 Project team for helping to develop and actively assisting the evaluation. The members of the team were:

L. F. Ennever, Director   1967–75
A. James, Deputy Director   1969–72
Miss S. J. Parker   1967–73
D. T. Radford   1967–72
R. Richards   1968–72
Mrs M. Horn   1971–72
Mrs A. M. Mattock, Secretary   1967–75

1 The origin, aims and methods of Science 5-13

Introduction

Science 5–13 has been one of the larger and longer-lived projects of its chief sponsor, the Schools Council. During the seven years of its activity it produced twenty-five books which aimed to provide various kinds of help in teaching science to children aged five to thirteen. One of the books provided an introduction to the project, discussed children's learning in relation to science and suggested objectives for children's science activities; seventeen books were units for teachers giving ideas of learning activities through which children at various stages of development might achieve these objectives; six were background or resource books for teachers, and one was a unit aimed at helping teachers become familiar with the project's ideas and materials. There were no books for children, since one of the project's tenets was that teachers could best help children to learn science by selecting experiences and activities appropriate to the abilities, and engaging the attention, of individual children.

The project began in September 1967, about eighteen months after discussions had begun between the Schools Council and the Nuffield Foundation about the possibility of setting up a jointly sponsored project in primary science. It was originally conceived as a continuation of the Nuffield Junior Science Project (January 1964–August 1966), with the Foundation contributing a fixed sum, but the continuing responsibility for support resting with the Council. Shortly after the start of the Science 5–13 Project the Scottish Education Department was welcomed as a third sponsor, contributing ten percent of the Schools Council grant in return for the participation of Scottish schools.

While sharing many of the educational convictions of the Nuffield Junior Science Project, the new team naturally wanted to form its own ideas about how it would try to help teachers, rather than continue the production of the kinds of materials developed by the earlier team. Being a second generation project it had the advantage of being able to study carefully the methods, materials and assessment of the Nuffield Junior Science Project as well as those of other science projects both in this country and abroad, and it made full use of this benefit. Soon, the new project developed its own line of approach

S. 5–13.—I*

to the solution of the problems it confronted, and it was appropriate that it should drop its initial name 'Junior Science Project', which linked it rather too firmly with the Nuffield Junior Science Project. The title 'Science 5–13' was chosen at the first meeting of the Project Consultative Committee.

Originally the project was sponsored for three years; subsequently it was four times extended for a total of a further five years. The Board of the (then) Institute of Education and the Senate of Bristol University accepted the grant from the sponsors and administered the project. Team members were temporary members of the University Institute of Education, which later became part of the School of Education. For five years of its life the project occupied a large house in the University precinct where it set up its own offices, duplicating equipment, workshop and storage facilities. The project was thus centralized, its members meeting each other daily; the consequences of this arrangement extended far beyond the domestic level to influence the project's methods and procedures. Frequent informal meetings between team members meant that ideas could be discussed and points argued in their early stages. It was not necessary to wait for people working miles apart to come together before talking over suggested approaches, making criticisms, or expressing opinions. To the team this seemed to give a unity of purpose to their work, and added to the development of a shared philosophy. It did not mean that differences among team members were eliminated, but such differences were revealed and accommodated rather than their discovery being left to chance.

What the project aimed to do

At the beginning, the broad lines of the project's terms of reference were laid down in the Schools Council memorandum proposing a joint Schools Council/Nuffield Foundation continuation project in primary science (Steering Committee A, Paper No. 28, April 1966—unpublished). The relevant paragraphs were numbers 7, 9, 10 and 11.

7 The main direction for the work of such a project is seen as extending the lines of development initiated by the current Nuffield project while paying particular attention to the needs of older junior pupils, and pupils in the early years of the secondary schools. The existing Nuffield team necessarily concentrated their efforts on the needs of infants and younger juniors; the needs of older pupils are now therefore the main concern. . . .

9 The principal aim of the project is seen as the identification and development, at appropriate levels, of topics or areas of science related to a framework of concepts appropriate to the age of the pupils. The aim of the development would be to assist teachers to help children, through discovery methods, to gain experience and understanding of the environment, and to develop their powers of thinking effectively about it.

10 Account will naturally have to be taken of the different needs of children of varied ability, according to their interests and aptitudes. Similarly, the question of supplementing, to some degree, the content of different environments for children in rural and town schools is one which will need attention.

11 This is likely to highlight another area of study, namely the best way of increasing the average primary school teacher's knowledge of modern science. The secondment of teachers for additional training, in present supply conditions, does not seem likely to be a remedy. The team will be encouraged to stimulate local experimentation to meet this need, perhaps through courses based in the teacher centres already set up in some areas. The team may also be able to consider how to advise colleges of education about the content of curriculum and general education courses which would equip teachers better to tackle science teaching in the primary school.

The team director and members were entirely free to find means of implementing these proposals. The methods eventually adopted were developed during the lifetime of the project; rather than being decided at the beginning and thereafter pursued without change, they emerged empirically, each step being taken after sensing reactions to the previous one. Because of this evolution of ideas and methods, the production of materials and their evaluation can be most easily followed if the events are described in chronological order. The following diary of the more important events in the project's history indicates such things as the frequency of meetings of various committees, the number of courses run, the dates of major decisions, publications and trials of materials.

Diary of the main events in the life of Science 5–13

1966

February	Discussions begin between representatives of the Nuffield Foundation and the Schools Council on the possibility of a joint continuation project in primary science.
March	The Nuffield Foundation agrees to contribute about £18 000 spread over the three years of a joint continuation project.
April	The Schools Council's staff produce a memorandum proposing to its relevant committees that a joint project in primary science should be supported.
May	Schools Council committees approve a proposal for the project, its grant being about £48 000 total for the three years. Total budget £48 000 (Schools Council) + £18 000 (Nuffield Foundation).

July	Mr L. F. Ennever formally invited to be Organizer (later designated Director) of the project.
	Bristol University formally approached to accept the grant and to house and administer the project.
November	The Board of the Institute of Education and the Senate of Bristol University approve the offer of the grant.
December	One-day conference of representatives from Nuffield Junior Science Project trial areas to discuss future arrangements for primary science.

1967

April	Project Director officially takes up appointment.
June	Project Secretary, Mrs A. Mattock, appointed (remained with the project throughout).
August	Project Evaluator, Mrs W. Harlen, appointed (half time).
September	Official commencement of project for three years. Two full time team members appointed: Miss S. J. Parker and Mr D. T. Radford.
November	Project Director visits elementary science projects in the United States for one month.

1968

January	Scottish Education Department contributes ten percent of Schools Council grant in return for participation in the project.
March	First meeting of Project Consultative Committee: project renamed 'Science 5–13'.
	Extended visit (three days) of members of Nuffield Maths Project for discussion with Science 5–13 team.
	Project team attends and contributes to School Council Conference on Junior Science
June	Supplementary grant approved by Schools Council: total grant becomes £86 000 (£62 000 Schools Council + £18 000 Nuffield Foundation + £6000 Scottish Education Department), for three years.
July	One week course for teachers and administrators from trial areas and other parts of the country.
September	Mr R. Richards appointed full time team member.
	Second meeting of consultative committee.
	Schools Council Newsletter, *Dialogue*, No. 1, publishes article on Science 5–13.

October   Meeting of project team and Schools Council representatives
with publishers interested in producing the project's materials.
November Project Director visits Bangkok and Malaysia.

1969

January   Mr A. James appointed full time team member (later designated
Deputy Director).
February Third meeting of consultative committee.
Project publishers selected: Macdonald Educational.
March   First meeting of trial area representatives.
Two team members visit schools in Scotland.
Three-day course run for overseas students at Bristol.
June   Project's first Newsletter published.
July   One week Science 5–13 course for teachers and administrators
–August   from trial areas to introduce first set of units.
September Three-day Science 5–13 course in Scotland organized by Scottish
Education Department.
October   First set of units published in trial form.
Trials of first set of units begin in twelve LEAs in England and
Wales, and in four areas in Scotland.
December Schools Council approves additional grant of approximately
£40 000 to extend the project for two years. (Schools Council
grant brought to total of £102 000)
Second meeting of trial area representatives.

1970

January   One week Science 5–13 course for teachers and administrators,
mainly from trial areas, to introduce second set of units.
Second set of units published in trial form.
Trials of second set of units begun in twelve LEAs in England and
Wales, and in four areas in Scotland.
February Fourth meeting of consultative committee.
March   End of trials of three units of first set.
April   One week Science 5–13 course mainly for College of Education
lecturers.
May   Evaluator becomes full time.
June   Fifth meeting of consultative committee.
July   Third meeting of trial area representatives.
Number of trial areas increased to nineteen in England and
Wales, eight in Scotland.
End of trials of second set of units and one unit of the first set.

August   Schools Council Newsletter, *Dialogue*, No. 6, publishes article
            on *Early Experiences* unit.
September   Three-day Science 5–13 Course in Scotland organized by the
            Scottish Education Department.
November   Sixth meeting of consultative committee.
            Fourth meeting of trial area representatives.

1971

January   Schools Council approves grant of about £10 000 for additional
            work, including the unit *Understanding Science 5–13*. (Total
            Schools Council grant £112 000).
            Mrs M. Horn, appointed half time team member, to work on
            Unit on *Plastics*, partly supported by a grant of £2000 from the
            Plastics Institute and industry.
            One week Science 5–13 course for teachers and administrators,
            mainly from trial areas, to introduce third set of units.
            Third set of units published in trial form.
            Trials of third set of units begin.
February   Second Science 5–13 Newsletter published.
March   Seventh meeting of consultative committee.
            Further one year extension of project proposed to Schools Council.
May   Schools Council approves grant of £11 000 to extend project to
            September 1973. (Total Schools Council grant brought to
            £123 100.)
June   Fifth meeting of trial area representatives.
July   End of trials of third set of units.
September Eighth meeting of consultative committee.
December   Sixth meeting of trial area representatives.

1972

March   Beginning of trials of fourth set of units.
April   First final publications appear.
May   Ninth meeting of consultative committee.
            Seventh meeting of trial area representatives.
            School Council Newsletter, Dialogue No. 11, publishes article
            on Science 5–13.
June   End of trials of fourth set of units.
July   Half time team member leaves project at end of contract.
August   Two full time team members leave project at end of contract.
September Deputy Director leaves project at end of contract.

| October | Tenth meeting of consultative committee. |
| December | Eighth meeting of trial area representatives. |

1973

January	Schools Council approves supplementary grant £7500. (Total grants now as follows: Schools Council £130 600 + Nuffield Foundation £18 000 + Scottish Education Department £10 340 + Plastics Institute £2000 = total of £160 940.)
March	Eleventh meeting of consultative committee. Evaluator leaves the project and becomes director of Schools Council's 'Progress in Learning Science'.
April	One week course for teachers, advisers, wardens and lecturers, organized by project, staffed partly by experienced teachers from trial areas.
June	Schools Council approves supplementary grant £6600. (Total grants now £167 540.) Combined meeting of consultative committee (eleventh) and trial area representatives (ninth). Setting up of working party on dissemination.
August	Last full time member of team leaves project at end of contract. Director and Secretary only remain in project.
December	First meeting of working party on dissemination.

1974 & 1975

Project work continues in hands of the Director and Secretary until September 1975 after further extension and supplementary grant of £9600.

The project's materials

The choice of type of material produced by Science 5–13 resulted from the philosophy adopted by the project. In the quotation from the Schools Council memorandum (given on pages 2 and 3) the project was exhorted 'to assist teachers help children learn science through discovery methods'. The many interpretations of 'discovery methods' range from the idea of meaning children exploring their environment with little adult interference to learning decribed as 'teacher-facilitated' to distinguish it from 'teacher-directed' learning. Somewhere between these extremes lies Science 5–13's interpretation—that discovery methods mean children being given opportunities, carefully provided by their teacher, to gain firsthand experience of living and non-living material around them, being encouraged to ask questions, find problems and seek answers for themselves, and being free to communicate what they feel and find in an appropriate way.

Discovery learning is supported by a large body of opinion (for example the Plowden Committee[1]), but it is not without its critics (for example Dearden[2,3]). Opposition is often based on the assumption that the teacher has no role to play in the learning process beyond that of providing materials and keeping the peace. It seems appropriate, then, to outline the intended role of the teacher in discovery learning, as envisaged by Science 5–13, since the materials were developed to give the teacher guidance and support in this role.

As has been mentioned in the introduction, the materials were directed towards teachers and not children; it was held that teachers were capable of being, and should be, responsible for thinking out and guiding the activities of their pupils. To maintain teachers' freedom and give them help, was considered important, so that classroom activities could follow and form the interests, and match the intellectual development, of the children. A central aspect of the teacher's role was conceived as being that of a guide to discovery. It was suggested in the materials that the main guidance would be through dialogue between the teacher and child or children. Discussing their problems with their teacher would help children to clarify their ideas, perhaps to see a different approach, perhaps to connect previous experience with a new situation. Guidance could also come in the form of physical objects or materials, selected and made available by the teacher to suggest an investigation or way of approaching a problem. Part of the teacher's role was also to provide motivation, arouse curiosity and give opportunity for handling and exploring things through which the children could learn.

The project's units provide suggestions for experiments or activities which could be carried out by children when their attention is engaged by a particular topic. The objectives which children could achieve through working on the unit's activities are made explicit, and alternative objectives are indicated for children at different stages of development. Each unit begins with ideas for 'starting points', intended to stimulate interest and provoke inquiry. In most cases the unit provides background information about the topic at the teacher's level, which she may lack or not be able to find easily in a suitable form; in two cases this background information is in a separate book. The units do not constitute a course; they are seen as sources of ideas and guidance which teachers can use in planning and carrying out programmes of work which they devise to suit the unique requirements of their own classes.

The nature of the material is being stressed since it is germane to the evaluation. The strategy of an evaluation must of course be appropriate to the curriculum material being tried, and procedures which are useful when new material is in the form of a course for pupils may not be so when the material is flexible and teacher-selected. It is important, throughout this report, to keep in mind the type of material produced by the project, because this was a major factor influencing the choice of evaluation procedures and instruments.

How the project and its materials developed

In the first year the team consisted of the Project Director, two full time writers, and the Project Evaluator, then working half time. From the beginning the team worked closely with teachers, frequently visiting classes, sometimes trying out and developing ideas with some of the children, often meeting groups of teachers at centres to discuss help they would like with science and possible forms in which it might be provided. The concept of units for teachers was gradually refined during this year and various early models were written. Each attempt was exposed to criticism by other team members, by teachers, by members of the project's consultative committee, and others patient enough to give their expert opinion.

A parallel and major activity during this first year was the clarification of what the project was hoping would be achieved as a result of its activity. Defining objectives in precise terms had not, in 1967, generally been given the attention by curriculum projects which it is now acknowledged to deserve. Certainly the Nuffield Junior Science Project had made only a general statement of what children might achieve through science activities. Perhaps it was not surprising, then, that the first responses to the evaluator's questions about what it was hoped children would derive from their science activities were along the lines of 'a liking for inquiry', 'a questioning mind', or simply 'enjoyment'. Together the team set about analysing what these broad intentions meant in terms of behaviour changes in children. In this process the evaluator asked the awkward questions: 'What does inquiry mean?' 'What do children get from inquiry?', 'Can the same outcome be expected for children at any stage?' and bullied the other team members until some draft lists of more detailed statements of objectives could be drawn up. These early lists of objectives, like the draft units, were discussed with teachers to ensure that they were realistic and attainable. While giving the team the benefit of their views, the teachers expressed a strong desire to have the aims and objectives made explicit in the project's material: it was evident that many teachers felt concern about the purpose of science in the early stages of schooling and some confusion about what kind of activity for young children was implied by the word 'science'. They realized that a statement of objectives for children learning science could be of help both in providing an operational definition of the meaning of early science and in indicating the contribution to the aims of education which science activities can make in the first eight years of schooling. In response to the teachers' reaction, the project's list of objectives, originally drawn up as an internal document to guide the writing of the team, became an integral part of the materials intended for teachers.

Further discussion of the formulation of the objectives follows later, in Chapter 2. For the moment it is sufficient to note that thinking about objectives and trying to frame them was a task which occupied a considerable amount of time, particlularly that of the evaluator, in the first year. Five different drafts were produced and several formats tried out. The work on the objectives was by no means finished in the first year; three major revisions followed, and the list did not reach its final published form until the end of the third year of the project.

During the second year of the project, the team was enlarged by two more full time writers. After the first year's examination of the problem in hand, and small scale trials to explore the feasibility of possible approaches, the second year saw effort concentrated into writing units and producing material for evaluating them in large scale trials. The trials of this first set of units took place at the beginning of the third year of the project, so the trial unit and evaluation material had to be ready well in advance. A one week course was held in the summer at the end of the second year to prepare trial class teachers and trial area administrators for using the units and evaluation material.

Although ideas from several team members might be incorporated in a unit, the responsibility for producing a given unit rested with one individual writer. Naturally, there were variations between writers in the way they produced a unit, but a general pattern can be discerned. A unit began as a skeleton—a list of ideas which the intending author wrote down about a topic which he or she felt would make a good unit and would enjoy developing. The skeleton was passed round the team and the potential of the topic discussed. Criteria applied to decide whether the topic would make a good unit emerged as the following (quoted from *With Objectives in Mind*,* page 37):

1. It must be attractive to both children and teachers.
2. The content area must be near to children; that is (a) it engages their attention; (b) it gives them opportunity to do something, to construct, to collect, to explore and find out; (c) it stimulates them to think for themselves and causes spontaneous discussion.
3. It must be realisable, given the circumstances of the school. This also means that it can be conducted in a variety of situations.
4. It must lend itself to development; that is, it must suggest interesting possibilities.
5. It must further the teacher's objectives for the children, and be seen likely to do so.
6. It must give the kind of help that the teacher needs; not only long-term help, as through pointing out realisable objectives, but short-term help with methods and apparatus.

* See Appendix A, p. 92.

Once work on the unit had been agreed, the author generally arranged to meet one or two groups of experienced teachers at nearby centres to discuss ideas for activities. These groups would meet three or four times and, between meetings, the teachers would try out the earliest suggestions of the author— perhaps add their own ideas—and then report back on relative success or failure of various activities and on the lines of development which had interested the children. The question of what the children were achieving would also be discussed so that the objectives of the unit were considered at an early stage. This expansion of early ideas about activities and objectives and accumulation of new ones enabled the writer to produce the first draft of the unit, which was reproduced in batches of a few dozen for distribution to team members and for use in small scale trials. At this point, team members commented on the first draft, and the author arranged for more teachers to try it and discuss their reactions. Suggested changes and additions were incorporated in the second draft, of which about two hundred copies would be produced. Comments on the second draft were invited from the project's consultative committee. If time allowed, further trials were carried out at this stage, but in later years it was generally necessary to use this second draft material for preparing teachers to use the unit in the full trials. The final printed trial material used in the full trials was similar in content to the second draft, with the addition of photographs and more illustrations.

The upper part of Fig. 3 (on page 22) represents the course of development of a unit, which will be mentioned again in describing the production of evaluation material. Perhaps it should be emphasized that the sequence outlined is an attempt to convey the idea of the pattern of unit production and would not necessarily be true in detail for every unit written; some writers preferred to put down their own suggestions in more detail before sharing them with teachers—each worked in the way he or she found suitable.

In the third year of the project the team membership remained at the level reached in the second year; later that year the evaluator became full time. By this point the Project Secretary needed the help of two full time typists to deal with the production of unit drafts and evaluation materials. The production of units continued at an increasing rate. A writer did not deal with one unit at a time but began another one when ideas and opportunity for developing it occurred. In this way, at least one team member was responsible for four units in various stages of planning, preparation or trial during this year. Since the extension of the project for a further two years had been confirmed, there was no necessity to curtail production.

Units were tried out in sets composed of those which could be brought to the stage of being ready for trials simultaneously. It may be useful to indicate here, in Fig. 1, the timing and duration of the trials of the four sets of units, because the intervals between the four trials determined how much could be learned from the evaluation of one set before the evaluation of another set was planned and undertaken.

Age of project	Date		Trials in progress
First year	1967–68		None
Second year	1968–69		None
Third year	1969	September	First set of units
	1970	January	Second set of units
	1970	April	
	1970	July	
Fourth year	1970	September	
	1971	January	
	1971	April	Third set of units
	1971	July	
Fifth year	1971	September	
	1972	January	
	1972	March	
	1972	June	Fourth set of units
Sixth year	1972–73 ⟶		Understanding Science 5–13
Seventh year	1973–74		None

Fig. 1   Timing and duration of the trials of the units⋆

During the project's fourth year the results of evaluation of the first and second sets of units had been analysed, and rewriting the trial versions for the final publications began. In addition to making the kinds of changes which are described later, on pages 45 to 47, 56 and 58, a major task was to gather photographs, good examples of children's work and other illustrations required for the unit. Each unit author was responsible for directing the artist to produce drawings that were required, selecting photographs, agreeing with the publisher's editor as to the final form of the text, and later for checking galley proofs and final layout—all good experience for those who had not published before, but exhausting and time consuming. The main work of the fourth year was the production of trial editions of the third set of units. The third was the largest set of units and the trials, run and evaluated in the second half of the year, were on a larger scale than any other of the project's trials.

The project team was at its largest throughout the fifth year following the appointment of a part time writer to work on a unit about plastics, for the development of which the Plastics Institute contributed a grant of £2000.

⋆ For titles of units in the four sets please see Appendix A on page 92.

As in the previous year, production of units for the fourth set of units had to take place concurrently with the revision of units of the third set after analysis of evaluation results. A major new task undertaken in this year was the development of the first part of the unit called *Understanding Science 5–13* aiming to provide information about the project and help teachers make effective use of the project materials.

The fifth year was the last of the project at full strength. In July and August, 1972 three full time writers and the part time writer departed and, for this reason, the production and trials of the fourth set of units had to be hurried, probably more than we would have wished. During the year a successful application was made to the Schools Council for a further one year's extension of the project, in an attenuated form.

In the sixth year the project's membership was reduced to the director, half time, one writer full time—mainly engaged on producing the second part of *Understanding Science 5–13,* and the evaluator—full time for half the year, writing reports on the evaluation. Apart from the individual responsibilities of the remaining members, the emphasis of the team's activities in this final year was upon dissemination of the project's materials and ideas. It became evident that the task of dissemination was a complex one about which very little was known. The concern of the Schools Council in this matter was shown by their setting up a working party to look into the problem of dissemination of projects' materials in general, and by granting Science 5–13 a further year of life to continue its work on dissemination.

The last full time writer left the project at the end of the sixth year and, during the seventh year, the director and secretary cooperated with Science 5–13's own working party on dissemination. The work of dissemination was continued into another year, the project's eighth year, with the director and secretary again as the sole members—but supported by a 'Science 5–13 Aftercare Committee'!

References

1 Central Advisory Council for Education (England) *Children and their Primary Schools* [The Plowden Report]. HMSO, 1967.

2 Dearden, R. F. 'Instruction and learning by discovery.' In R. S. Peters (ed), *The Concept of Education.* Routledge and Kegan Paul, London, 1967.

3 Dearden, R. F. *The Philosophy of Primary Education.* Routledge and Kegan Paul, London, 1967.

2 The role and purpose of evaluation in the project

The role of the evaluation

Evaluation in this project has been concerned throughout to assist in the production of the units and other books. Its role can be described as essentially 'formative', or 'ongoing', which indicates that it takes place during the development of material and is intended to improve trial versions before final publication. Evaluation can take another role in which it is used to investigate the effectiveness of final published versions, when it is described as 'summative' or 'terminal' evaluation. Sometimes it is difficult to designate an evaluation as being either wholly formative or wholly summative since a formative evaluation often produces information useful in a summative context, whilst the summative evaluation of the products of one curriculum project could be the first step in the formative evaluation of the next wave of curriculum development in that field. In the case of Science 5–13 some useful summative information was indeed produced by the evaluation but nevertheless its role was intentionally formative.

Thus, the evaluation took part in assisting the development of the project's ideas and materials, but the form of assistance it gave was of course different from that contributed by other members of the team. The evaluator deliberately avoided being directly involved in producing ideas for activities or writing the units; instead she was concerned with clarifying the objectives of the project and providing information as to how well the materials were achieving their intended purposes. The roles of writing members and evaluator were therefore distinct and yet interconnected. Evaluation was an integral part of the project's development from the start, but was separated from other functions of the project by being the sole responsibility of one person.

There are advantages and disadvantages to this close relationship between evaluator and other team members but, on the whole, the advantages—which centre around ease of communication—outweigh the disadvantages, which arise from loss of objectivity on the part of the evaluator. It is no doubt important for the evaluation of material to be in the hands of someone who is not committed to it, emotionally or intellectually. Such commitment is bound to follow in anyone who has wrestled with the problems of developing and writing material, and it is evident that the person thus committed is less likely to be able to stand back from the material, look at it objectively, and gather evidence about it impartially.

It is not a fault in the writer, but a result of his necessary involvement in his work, that what he produces in more likely to be improved if it is evaluated by another person. But this is only so if this other person is sufficiently well informed about the purpose, content and context of the work to enable him to gather useful evidence. The uncommitted evaluator is faced with the problem of communication—if he tries to remain impartial and objective he may also be ignorant of many things which he should understand to do his job properly. It is no easy matter for an evaluator who is not close to the project team to understand the aims of their work sufficiently to plan a valid and useful evaluation. This is particularly so in the case of formative evaluation, but also true to some extent in the case of summative evaluation. When the evaluator is a team member the communication problem is reduced, but at the risk of too much involvement in the development of material. So, there are drawbacks to both situations, which cannot be avoided but only minimized by being fully aware of them and consciously taking the necessary steps to reduce them.

There were advantages arising from having made a commitment to formative evaluation right from the beginning which went beyond having information fed back as to how effective the trial material was in practice. Without going into detail on this subject (an extended account was given in a paper by the evaluator published in 1971*), it is worth noting that it was felt beneficial —even if uncomfortable—to have someone in the team who had to ask needling questions. To prepare test items for children or questionnaire items for teachers meant probing the material thoroughly, digging down to find what children were intended to achieve from the activities, and how teachers were expected to guide their children. During the development of evaluation material, writers could not avoid facing these crucial questions, nor could the evaluator avoid coming to grips with the essential purpose of the material. One result was a greater awareness on the part both of writers and evaluator of the others' problems and a sharpening of thought among all team members about the task they were jointly undertaking.

Having an evaluator as a team member also played a large part in ensuring that the project's objectives were stated explicitly and in some detail, rather than remaining implicit, as was the case for many curriculum projects up to that time. Since helping to work out Science 5–13 objectives was the first of the evaluator's activities it is appropriate now to take up this story again.

* 'Some practical points in favour of curriculum evaluation' (See Appendix C, p.99).

Helping to frame the objectives of the project

It has been mentioned above that clarifying the objectives of the project was a concern of the evaluator and occupied a large part of her time during the project's first year. The draft list of objectives which was discussed at many different meetings with teachers went through many cycles of revision during which problems relating to how specific the objectives should be, how precisely worded, whether expressed as processes or outcomes, etc., were encountered.

The first tentative list had its origins in an attempt to find the 'framework of concepts appropriate to the age of the pupils', referred to in the Schools Council's proposals for the project (see pages 2–3). Such a framework did not exist but, drawing on the findings of Piaget and direct experience of investigating the development of children's scientific concepts, the evaluator produced a chart of the scientific concepts and knowledge indicating what might be expected of children at different stages of mental development. This was put up for the rest of the team to use for target practice. Bringing their knowledge and experience to the matter produced a considerably changed document, which was the first to bear any relationship to the eventual statement of objectives. In it were such items as 'Recognition of broad classes of materials, e.g. plastics, metals', 'Examination of a wide range of materials', and 'Tabulating information and using tables'. There were about ninety-five items, each relating to one of three stages of development:

Stage 1. The transition from intuitive to concrete operational thinking and the early phases of concrete operational thinking.
Stage 2. The later phases of concrete operational thinking.
Stage 3. The transition from concrete operational to formal operational thinking.

As well as being grouped according to the developmental stage which seemed appropriate, they were arranged for convenience under headings of more general statements. During the many revisions the number of these headings was changed and their wording was altered so that they expressed the broad aims to which the more specific statements below them were related. In turn, these broad aims were synthesized in a statement of overall aim: 'Developing an inquiring mind and a scientific approach to problems.' So the aims at various levels of generality were related to each other rather as members of a family in a family tree, as can be seen in Fig. 2.

Later it became evident that the most helpful way in which this relationship between statements at different levels could be conveyed was by describing it as if the more specific ones were derived by analysis from the more general. In fact the process was not so straightforward, but the rationalization was justified by the more logical and easily grasped picture it conveyed.

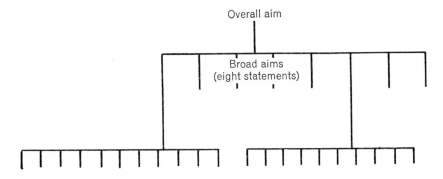

Fig. 2   Relationship between broad aims and objectives

The statement of the general aim and the broad aims alongside the more detailed statements has been found to have considerable value. Broad aims on their own are of little use since they provide no guidance for day-to-day working in the classroom, but so also are specific statements on their own of limited value in another sense. The broad aims should not be forgotten because they are, in the end, the overall purpose of learning experiences. While pointing out the trees it is useful to keep the wood in view. The achievement of specific objectives must somehow add up to the achievement of the broad aims, and the process of integrating the parts to make the whole is no more automatic than the process of analysis which can, in theory, be used to turn broad aims into specific objectives. It was found helpful to teachers to have not only the specific statements pointed out but also the broad aims and the overall goal as well. In this way, links between aims and objectives at one level and at another can be kept in mind, and it becomes more possible to see where achievement of a particular skill or ability fits into a wider pattern of achievement instead of seeming an end in itself.

It soon became apparent that, in order to express as unequivocally as possible what was meant by each of the statements, we would have to decide how achievement of the skill or ability or concept in question would affect a child. We had, in other words, to express the objectives in terms of changes in children's behaviours. This course of action was also urged by curriculum experts whom we consulted in the early stages of defining the objectives. Some authors on the subject of curriculum studies (for instance Wheeler[1]) state uncompromisingly that all statements of goals or objectives must be expressed as behaviour changes. Others (Eggleston & Kerr,[2] for example) recognize that, while it might be desirable to aim at defining goals and objectives in behavioural terms, it may not be as easy in some subjects as in others. Further discussion of the pros and cons of expressing objectives to different degrees of precision is not relevant here but can be found elsewhere (Harlen[3]).

Even a quick glance at the latest version of Science 5–13 objectives shows that not all are stated strictly in behavioural terms. Successive revisions brought them closer to specifying behaviour changes but progress towards a complete behavioural description was limited by an important consideration: the use being made of the objectives. They constitute a working document, being used by teachers, many of whom heard the word 'objective' in the context of education for the first time on being introduced to Science 5–13 material. Closer behavioural specification tends to introduce jargon and lead to longer and more complex statements—all things which would deter teachers from using the objectives. So, at that time, it seemed best to keep to a terminology which communicates with teachers, providing objectives they can use, even though it meant peppering the list with words such as 'awareness', 'knowledge', 'appreciation', which are frowned upon by purists. Our point of view may be better appreciated in the light of this critical comment on Science 5–13 objectives from a Bristol infants' teacher:

It seems to me that the objectives tend towards the clinical. 'Appreciation of the variety of living things' does not stress the beauty and glory of the world, perhaps the aesthetic side comes automatically to some children but others can look at a horse chestnut in flower and write down what colour blossom it has etc., without realising the beauty of the whole and this is because they have been told to 'dissect' it.
Perhaps an objective could be 'encouragement just to wonder'. Instead of going on a walk to collect wild flowers it might be interesting just to walk and look at everything (not at the ground and how many things your friend has collected), and soak in the glory of nature.
Another objective might be to notice the variety of movement, appreciation of machine kinetics being an example. The grabbing mechanical digger, the tall graceful cranes swinging their loads, the halting lorry with air brakes, in heavy traffic, the heaviness of the cumbersome ocean liner and quickness of the tugs, the steam roller, pneumatic drill and pneumatic hammer. All these things have their own different rhythm and movement.

Quite evidently this teacher has a very keen grasp of what she wants her children to achieve. She is intuitively working towards objectives which are very clear in her own mind. But, in her opinion, to analyse them would be to lose sight of them; far better to leave them alone.

The purpose of the evaluation

It is useful to distinguish between the *role* of the evaluation, the part it played in developing the material, and its *purpose*, that is, the questions it attempted to answer. Throughout the project the role of evaluation was the same, but certain changes can be discerned in the questions it addressed. In planning the evaluation of each set of units it was necessary to decide which major questions we should attempt to answer before the means of finding answers could be devised. For the evaluation of the first set of units, being planned in 1968–69, the most important questions to ask seemed to be 'How much does the material help the children achieve the stated objectives of the units ?' and 'What changes should be made in the material to make it more helpful ?' After interpreting and using the results from the first set evaluation it became clear that information which was more effective for formative evaluation would come from changing the first of these questions to: 'How well does the material help teachers to provide intended learning experiences and interact with the children according to the project's intentions ?' Reasons for this change in focus of the evaluation emerge in later chapters.

Before any decisions were reached about procedures for finding answers to these questions certain guidelines, which were to direct the decisions, emerged from considering the type, and intended use, of the trial material and the situations in which it was to be tried out. The first of these guidelines related to the complex nature of behaviours which the material was hoping to affect and the difficulty of obtaining valid measurements of changes in behaviour. It would be necessary to use a variety of methods of assessment and to gather evidence about any changes in attitudes as well as in cognitive abilities. Tests or other instruments used in formative evaluation have to be prepared quickly and there is no time for the luxury of refining them through repeated pilot trials. The evaluator has therefore to accept that the instruments are less reliable than might be possible if more time for their refinement were available. Low reliability generally means lower validity, so the results of any one test instrument may not in isolation be very useful. However, if two or more instruments of the same or different type were used, and the results tend to support each other, then more faith can be put in the results.

The second guideline was the importance of choosing or devising evaluation instruments which would be appropriate to the behaviours being investigated and would interfere as little as possible with the expression of these behaviours. A project's material is invariably intended to be stimulating and interesting for both teachers and pupils and, if the hoped for positive response emerges, it is essential that the evaluation procedures should not diminish it. The evaluation techniques should be as enjoyable as possible—if they cannot be unobtrusive—but in any case efforts should be made to minimize the risk of detrimental interference.

A third point was related to the quantity of information which might be gathered. Collecting evaluation data involves other people, mainly the teachers and children, whose time and efforts have to be respected. In the enthusiasm for collecting as many data as possible to give a many-sided description of the trials, it is important to consider whether in fact all the information could be used. Time is again a factor which has to be considered; it limits the amount of data which can be handled in formative evaluation. It would be unfair to teachers and children to ask them to participate in the data gathering, and then disregard their results because too much information had been gathered. There is, therefore, an obligation to think ahead in planning the evaluation of trials and to collect no more information than can be handled in the time available for analysis and interpretation.

A fourth guiding principle was that it would be beneficial on several counts to involve teachers in gathering information. It would benefit the evaluation if teachers carried out any testing of their children, since this would decrease the interference with normal work caused by the arrival of an unfamiliar test administrator. It would benefit the evaluation for trial teachers to meet together to discuss the purpose of the evaluation, the meaning of some of the questions posed on the forms they are asked to fill in, and the problems encountered in the trial work which may not be covered by points in the questionnaires. But the greatest benefit would come if teachers' anxieties about evaluation could be dispelled; this seems very much more likely if they are thoroughly involved in the process than if they are peripheral to it.

The problem of attitude to evaluation was one the project had to take very seriously. The idea of evaluation creates a considerable amount of suspicion among teachers and others involved in primary education in this country. To most the word 'evaluation' is taken to be synonymous with 'testing'. In primary schools generally there is an understandable dislike of testing, a dislike arising from the experience of the effect of 'eleven-plus' selection examinations. Teachers who sighed with relief at seeing the end of these examinations, which had a restricting effect on the curriculum, and induced anxiety in children, teachers and parents, are very naturally unwilling to accept with favour what may appear to them as more tests disguised under the name 'evaluation'. There are also other historically determined prejudices, dating back to the last century and transmitted from one generation of teachers to another. It is very understandable that, until there is more general appreciation of the role of evaluation in education, some should think of it as putting the clock back.

So, an evaluator may well find in the primary school a situation in which tests are unwelcome and are not a normal part of the work. The problem is not the same at the secondary level, where examinations, both internal and external, are normal occurrences. It is not a trivial point, which merely causes discomfort and calls for tact; it is extremely important in the planning of tests as part of an evaluation programme. It means that, if tests of the conventional kind are used in the primary school, then—because they are not a normal part of the work—they create a distortion of the pattern of work and an artificial situation in which an 'educational uncertainty principle' operates—the interference to the work caused by evaluating it means that nothing is learned about the work without such interference.

The project has tried to tackle this problem by doing as much as possible to inform teachers about the reason for evaluation and the plans devised to carry it out, and by enlisting the help of teachers with the evaluation. It was our experience that this approach met with success; when teachers have had a chance to appreciate that evaluation is important for the development of the material, that it was concerned with the material and was not assessing their children or themselves and that they had a valuable part to play in it, then they were prepared to cooperate without prejudice. We tried to make them feel genuine partners in the project's attempt to help their work. It was envisaged that involvement in the evaluation would benefit teachers by making it evident that achievements other than knowledge can be assessed and that attempting to assess these things can encourage a critical appraisal of their own work.

Producing the evaluation material

The four principles outlined above guided the choice of procedures and instruments to be used for the first evaluation of the first two sets of units; experience of these first trials suggested shifts of emphasis but no fundamental change in the principles when they were applied to the later trials. Most of the discussion of the decisions taken about evaluation materials is to be found in the following chapters, but it is convenient to anticipate the plan for evaluating the first set of units so as to illustrate the way in which unit and corresponding evaluation materials were developed in parallel.

A prominent aspect of the evaluation of the first set of units was measurement of changes in children's achievement in relation to the objectives of the units. Of course there were no ready made tests for the units' objectives, so these had to be devised. Different kinds of information about the trial material were gathered by means of three separate questionnaires which were filled in by the teacher and a further form completed by a team member after visiting each trial class.

In Fig. 3 on page 22 the parallel development of the unit and the evaluation material is represented. The stages through which a unit typically passed during its writing (described on pages 10 and 11), are indicated along the upper arm of the diagram. Along the lower arm are the stages of devising the evaluation material. The planning of the evaluation began as soon as the first complete draft of the unit was produced. Of the instruments to be used in the evaluation, the tests for children were the most elaborate and time consuming to produce, so they were the first to receive attention. The first draft of some possible test items was discussed with the author of the unit before any test production was begun. These discussions were of considerable

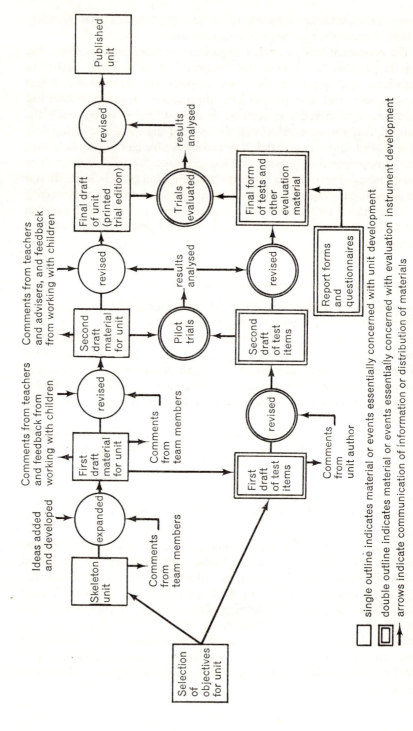

Fig. 3 Development of unit and evaluation material—first set of units

benefit to both participants, as part of the close communication which it was possible to establish in this project. Each found value in seeing how the other had interpreted the unit's objectives and, without necessarily having any ideas or opinions changed, appreciated another point of view. The chief reasons for discussing the test items at this early stage were to ensure as far as possible that the items would detect the kind of behaviour change which the unit's author was intending to promote, were acceptable in terms of their content, and covered adequately any aspect of the unit which the author particularly wished to be explored. To do this involved digging deeply into the meaning of the objectives, analysing situations in which achievement could be promoted or could show itself, giving consideration to the degree of achievement expected, and so on. Such exchanges, early on—before ideas had become too firmly rooted—caused both writer and evaluator to think in more depth about the purpose and effectiveness of what they were producing.

After this, a test was produced in time for use with some of the pilot trials. The films and booklets for the test (see pages 27 and 28) were made as carefully as if they were the final versions, but only reproduced in limited quantities (about 1000 test booklets and two copies of the films). The test was then given to about six hundred children to provide data for item analysis. Some of these children were involved in pilot work with the unit and, where this was so—and it was convenient to test the children again after the trials—these pre- and post-test results were used as a runthrough of the techniques which would be employed later in the full trials. The pilot tests were administered by the evaluator, except in the case of a small group of schools in Croydon where there was a feasibility study of the practicability of teachers giving the tests themselves.

The final form of the tests was decided upon as soon as the final draft of the unit was ready to be printed. Thus, any last minute changes in the unit which might have consequences for the test material could be accommodated. At this point also the questionnaires and report forms, which were to give other kinds of data for evaluation, were written.

References

1 Wheeler, D. K. *Curriculum Process*. ULP, London, 1967.
2 Eggleston, J. F. and Kerr, J. F. *Studies in Assessment*. EUP, London, 1969.
3 Harlen, W. 'Formulating objectives—problems and approaches', *British Journal of Educational Technology*, **3**, 3, 1972, 223–36.

3 Evaluating trials of the first set of units

Deciding what to evaluate

The plan for evaluating the first units to be produced by the project was decided as a result of bringing together three considerations: the role of the evaluation, the guiding principles outlined on pages 19 and 20, and the kind of curriculum material being developed. The role of the evaluation was to gather information useful in improving the trial units; the guiding principles summarized the main factors that should be taken into account in deciding what and how much information to gather, and how it should be gathered; the nature of the units determined not only the content of the evaluation material but also the scope of aspects it should attempt to investigate.

At the time of planning the first evaluation trials (1968–69), there were no examples of similar work to learn from and, although a considerable amount had been written about the theory of formative evaluation, most of this was relevant to the development of material in the form of a course. In evaluating such course material, emphasis is laid upon testing achievement of objectives because this is considered the all important criterion of success or failure of the material. The ultimate aim of all educational material is to produce change in the behaviour of children. The changes intended to be produced by the Science 5–13 units were expressed in the project's 'objectives for children learning science'. Yet the Science 5–13 materials were written for teachers and not for children; they affected the children only through the agency of the teachers. The teacher mediated between the materials and the children, taking a part which was intentionally far from being a passive one, but was instead very active and creative.

There were two important links in the chain from the project's materials to the children, which can represented diagrammatically as follows:

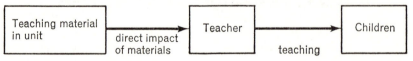

If the direct impact on the teachers of the units, plus courses and other aids to understanding the materials, was not effective in communicating the project's ideas, then the units would fail to produce any results in practice—no matter how good in theory the ideas may have been. On the other hand, if the units communicated the ideas well but the material was not effective in helping children learn, again there would be no practical result, but for a different reason. In addition, the mediation of a particular teacher in using the materials would be influenced by her personal opinion of the units and the extent to which the class and school environment allowed her to use the materials along the lines intended.

Realizing the complexity of different factors which could influence the outcomes of using the materials naturally led to the conclusion that it would be pointless to measure only the outcomes. To explore relationships between certain circumstances and certain findings would require investigation both of links in the chain between the unit and the children, and of the influence of the learning environment as well. This environment included the social as well as the physical, both within and outside the school.

It seemed as if a multitude of tests, observations, questionnaires and forms would be required, but practical considerations limited the choice. Whatever information was gathered had to be gathered quickly and interpreted quickly, otherwise it would be no use for its purpose of helping with the revision of the units. We have already mentioned that time is against the use of elaborate and carefully refined instruments for formative evaluation. So it is all the more important to make the instruments as valid as possible, that is, carefully designed to detect what they are intended to detect and not something else. The guiding principles were also borne in mind, particularly with regard to choosing means of finding information which would not interfere too much with the work in progress. Finally the selection of evaluation instruments had to be such that together they provided a suitable balance of different kinds of information.

Eventually four aspects of the trials were chosen for investigation: changes in children's behaviour; the opinions and reports of teachers; the interaction of teacher, children and material, and the learning environment. Figure 4 indicates the techniques which were used for investigating each of these four aspects.

The evaluation materials

Very little needs to be said about the questionnaires and report forms, Teachers taking part in the trials were asked to complete three forms:

Form A   mainly blank paper on which teachers were invited to write an account of relevant activities of the children during the trial period;

Form B   a questionnaire, mostly pre-coded, asking for information about the class and school, relevant biographical data about

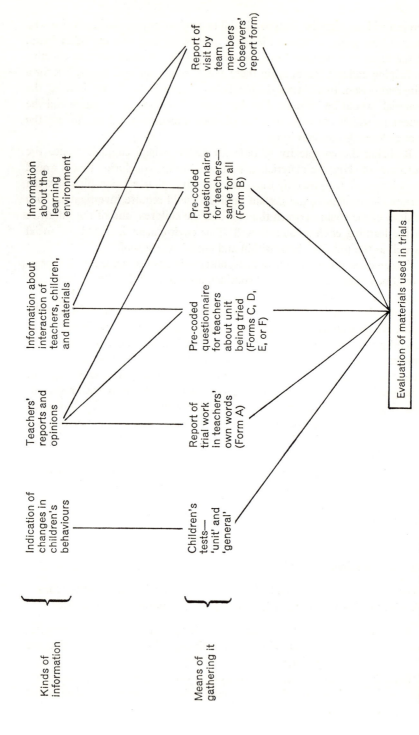

Fig. 4 Information gathered in evaluation of the first set of units

the teacher, and opinions of the book *With Objectives in Mind*;

Form C, questionnaires, mostly pre-coded but with provision for free
D, responses as well, asking for information, opinions and
E, suggestions about the unit. (One of these forms was completed
or F according to which unit was being tried by the teacher.)

In planning these forms an attempt was made to arrange the questions in the order in which things might be encountered, so as to encourage teachers to record their reactions at frequent intervals rather than leaving all the recording to the end of the trials. In addition to the written comments collected on the forms, teachers' opinions were also collected informally at meetings and during visits by team members to the trial classes. All teachers taking part in the trials were provided with a copy of *With Objectives in Mind*, the introduction to the project.

Information about interactions in the classroom was recorded after informal observations during visits by team members. It had not proved possible to arrange for the prolonged observation and recording in the classroom, normally required if formal analysis of interaction is to be attempted. Instead, each trial class was visited by a team member while work was in progress and straight after the visit the team member filled in a form, recording what had been observed about such things as the class organization, the use being made of the unit, the teacher/pupil relationship, how the teacher had seemed to react in discussion to the project's ideas, etc. To share the work of class visiting, each of the four full time team members, excluding the director, was made responsible for visiting schools in three areas. Where possible, these were areas in which units written by the visiting team members were being tried. In Scotland class visits were carried out by HMIs.

Although the tests for the children were no more important than the forms and questionnaires, more must be said about them as they were not the conventional kind of test. Since the objectives for learning science in the early school years are concerned with how a child will act or deal with practical problems, the most obvious way to find out whether he has or has not developed a concept or idea is to present him with a real situation, chosen so that he needs to have developed the concept or grasped the idea in order to deal with the situation successfully. However, since this would require individual administration, it was completely impracticable in the present case. The most attractive alternative, which would avoid too much dependence on verbal ability and some of the other disadvatages of the pencil and paper form of test, was to use moving film. Although a two-dimensional medium, the movement of objects and apparatus provided the necessary cues for their real nature and form to be appreciated. The most practical form was chosen; to use a projector suitable for use in daylight—one with integral rear projection screen, for convenience. This kind of projector has the added advantage of resemblance to a television set, already familiar to children as a source of information.

The test items were written to detect the various kinds of behaviour described in the objectives. Only a minor fraction of the total number of the project's objectives relate to 'knowledge'; by far the majority deal with the many other different kinds of behaviour involved in learning science. The old idea that only knowledge can be tested has been largely dispelled, but it is still true that knowledge is the easiest to test by conventional means. Fortunately the film medium is a good one for presenting test items which do not merely test knowledge. Film may be used to present a problem, the solution of which calls, for instance, for ability to identify and manipulate variables, in a way which is clearer than a verbal description of the problem. It is one way of introducing problems which test understanding of such things as sorting, grouping and classification. Data can be communicated through film easily for items in which all information is provided and the question demands reasoning or explanations. After watching the film sequence for a particular item, the children recorded their answers in booklets by ticking one of the alternative responses given.

In devising items, situations or problems were thought up such that it was possible for children to show by their responses that they had achieved a particular objective. These situations or problems were made the basis of the items. In choosing them, care was taken where possible to avoid using problems which were exactly the same as those described in the units. The achievement of objectives was the criterion of success in this part of the evaluation, not how thoroughly the problems suggested in the unit had been studied. To have based the items directly on problems selected from the activities suggested in the unit would not only have implied that those activities were more worthy of attention than others, but would also have indicated that the teachers were expected to follow the unit in a way which was not intended at all. As has already been said, the units are not a course; therefore, a test based on a course had no function in this situation.

The four units in the first set were all related to Stages 1 and 2 and were to be given trials with children aged seven to twelve. The units were all written in a similar pattern; each took a selection of about ten objectives from the general list and showed how these objectives might be achieved through activities in the particular subject area of the unit. Activities relating to Stage 1 objectives were in earlier chapters of the units and activities relating to Stage 2 objectives were in later chapters. Within the chapters, the objectives were taken one by one and activities suggested through which it would be possible for children to achieve each of them.

It followed that a separate test was required for each unit, so four tests, called 'unit' tests, each consisting of thirty-four items were produced. But, in addition, there were objectives common to all units: objectives chosen to guide the approach to active learning in science pervading all the project's materials. These were called the 'general' objectives, and a 'general' test was devised to assess their achievement. In this test the items tested application and transfer to other problems to an even greater degree than the unit test items. For example, in one item the film shows two pieces of paper, equal in size —one plain writing paper and the other newspaper. They are picked up and a strip is cut from each; the strip of newspaper is wider than the plain paper but they are the same length. These strips are then attached to separate bulldog clips hanging from a frame. Similar clips are put on the bottom of the strips and weights hung from them as the first part of the sequence ends. At this point the commentator, administering the test, asks the children to record in their booklets whether or not they think this is a fair way to test which kind of paper is the stronger. The film sequence then continues; two more strips of paper are seen hung side by side on the frame. This time they are of equal width as well as length, and again weights are being hung from them as the sequence ends. The children are then asked to record in their booklet their judgement of this test of the paper, by ticking 'fair' or 'not fair' in the second part of the question. The objective under test here is 'awareness that more than one variable may be involved in a particular change'. (Stage 1).

The general test was in two parts, the first of items concerned with the 'general' cognitive objectives, and the second concerned with the objectives of the affective domain. In the latter, an attempt was made to measure children's liking for certain activities, both those involved in science and in other areas of work. The items asked the children to state their liking for the various activities in terms of a three-point scale presented to them visually. In the introduction to this part of the test the film sequence showed two children eating ice-cream. The children taking the test were asked to imagine themselves doing this and say how much they liked doing it. It was supposed that almost all would like this 'very much', so a personal standard for liking something 'very much' was established. The second sequence showed the children polishing shoes—something which provided a standard for liking 'not very much'. After this the sequences showed other activities and in each case the children were invited to imagine themselves doing the same things and to note how much they liked it. The point of filming these activities rather than simply giving a list for the children to assess was that titles such as 'looking after animals' or 'gardening' could be interpreted in many different ways by different children, whereas the film showed a definite activity. The children were thus all responding to the same situation not to their own idea of what the activities involved.

In the last chapter we described how the development of these evaluation instruments went on alongside the development of the unit material. The production of tests for the first two units to be written, *Metals* and *Working with Wood*, began a whole year before the trials began. There was, therefore, time to give the tests for these two units thorough pilot trials. Results of pilot tests administered to about six hundred children, spread throughout the seven-to-twelve age range, were used for item analysis. Analyses of all responses, right and wrong, with the help of the university computer, made possible an economical revision of the tests, since poor items could be improved rather than eliminated. Following the analysis, the films were remade and new booklets produced for the children, in sufficient quantities for the full scale trials.

The making and remaking of the film loops were arduous and time consuming, but the detailed probing of what an item was supposed to be testing and how the problem should be presented had a beneficial backwash on the thinking that went into a unit, as has been mentioned on page 23. The filming was often fun too. Our technique improved, though the films were never of high quality. One team member, Sheila Parker, helped to produce the films by manipulating things or carrying out the actions shown on the film. Sometimes the action was out of doors, as in the case of many items for the unit, *Trees*. It will be a long time before Sheila forgets the sequence where she had to run under the shelter of a tree, and kept running for the sake of the film, despite losing her sandals in mud about two feet deep. We only ever found one sandal; the Project Director allowed replacement under the heading of 'photographic materials'!!

Organizing the trials

Because of its history, the project inherited arrangements for trial areas set up by its predecessor, the Nuffield Junior Science Project. The trials were to be carried out in twelve LEAs in England and Wales which had been chosen as pilot areas for the earlier project and in four areas in Scotland. One advantage of conducting trials in the same areas as the previous project was that, in these areas, it would be easy to find classes which had already had experience of active discovery methods and in which, it was hoped, some learning through discovery was already going on. These were the kind of classes wanted for the trials so that the value and effect of the materials would be evaluated without the influence of introducing a new way of working. A disadvantage was their geographical spread, from Kent to Anglesey and Yorkshire and, of course, Scotland. In the long run, however, despite the problems of travelling for the team members, the wide distribution of areas turned out to be an advantage. Trial areas, after five years working with the project's materials, became centres of experience and good practice in Science 5–13 and were growth points for dissemination when the project completed publication. But this is jumping ahead several years; we shall return to this point in a later chapter.

A considerable amount of information was required about each class taking part in the trials. To gather a lot of information about a relatively small number of classes was considered far more helpful than to spread the information gathering over a large number of classes, but learn less about what was going on in each. The number of trial classes for each unit was chosen to be eighteen, and these were provided jointly from three pilot areas so each area provided six classes on average. Other classes in the trial areas and in other LEAs were able to obtain the trial units and were free to use them, but no feedback from them was requested; indeed, it was discouraged. The reason for this was that a single kind of information would not be useful without support from other types of information, and it was not justifiable to ask teachers to write reports which could not be used.

The children's tests were used to detect any changes in behaviour relating to the objectives. Measurement of level of attainment was necessary to measure change in attainment, but was not in itself of interest. Neither was there any interest in discriminating between individual children or individual classes, but only in the overall effect of the trial work. To detect changes during the trial period a 'before and after' pattern of testing was planned. It would not, of course, necessarily follow that the trial work had anything to do with any changes observed in this way, so a control group of children was tested in the same manner as the group undertaking the trial work. The function of the control group was to enable an estimate to be made of what part of the trial group's change in behaviour was the result of experiences connected with the trial material and what part of this change was due to the combined effect of test sophistication, maturation and other experiences.

The control group would only serve this function adequately if carefully selected to match the trial group as fully as possible. Complete matching on a large scale is impossible, so in this case differences between classes in the two groups was randomized. The mechanism for this was as follows: pairs of classes were suggested for participation in the trials, the classes in each pair being as similar as possible in respect of such things as age range of children in them, type and size of school, characteristics of neighbourhood, experience of teacher, etc. These classes were then assigned randomly to either the control or trial group, so that no bias was introduced into the system.

Selection of trial and control classes was carried out with the help of representatives from the trial areas. Each area nominated a person, usually a local inspector, science adviser, or teachers' centre warden, to be responsible for Science 5–13 work in the LEA and for liaison with the project team. The area represenatives first met together with the team in late March 1969, while the planning was being undertaken for trials to begin later that year. At the first meeting the general purpose and design of the evaluation was explained and the representatives had a chance to examine and discuss trial units and draft evaluation material. As the teachers' questionnaires were then in an early stage of development, the comments received at the meeting and subsequently were helpful in deciding the final form they were to take. Criteria for selecting classes were explained and the representatives were asked to choose six pairs of classes, to send relevant details of these classes to the project on forms provided, and to leave the selection as to which of each pair would be trial and which control class to be made randomly by the project. In choosing the classes the representatives were asked to keep in mind whether the teachers would be able, if chosen as trial teachers, to attend a one-week Science 5–13 course to be held in the summer holiday, whether they were likely to leave their school, or become promoted, or pregnant, during the trial period.

An attempt was made to find pairs of classes which could represent different age groups in the junior school and different types of school. It was not thought necessary to make any finer distinction by age than two groups, seven-to-nine and ten-to-twelve. Similarly, a simplification was made in sampling different types of school by selecting from three broad bands: rural, prosperous urban, and disadvantaged urban. Thus, there were two age groups from each of three types of school to be included, giving six kinds of class. It was not to be expected that each of these could be sampled within any one area, but the three areas trying out each unit tried between them to find the sample of eighteen classes which included each kind. The sample aimed for in the case of each unit was as shown in Table 1.

Table 1   Sample for evaluation of one unit
(provided jointly by three pilot areas)

| Age range | Type of school | Number of classes | |
		Trial	Control
7–9	Rural	3	3
10–12	Rural	3	3
7–9	Urban (prosperous catchment area)	3	3
10–12	Urban (prosperous catchment area	3	3
7–9	Urban (disadvantaged catchment area)	3	3
10–12	Urban (disadvantaged catchment area)	3	3

At first the names of some 'substitute' classes were requested to 'fill in' if selected classes dropped out but, once the trials had begun, it was not practicable to substitute classes; fortunately it was hardly necessary. The careful choice of classes on the part of the area representatives meant that only three out of a total seventy-nine trial classes dropped out during the trials.

By the time the first meeting of trial area representatives took place the decision had been made to ask teachers to administer the tests to the children. The alternative would have been for one person in each area to pass from school to school with the films, projector and booklets. Such an arrangement would have had the advantage of standardizing the mode of administration but would have had many disadvantages. A visitor to the school, however friendly, would give the testing an atmosphere of unfamiliarity, possibly anxiety; he would not know the children, how quickly they could work or whether they were following his commentary to the film. Neither could he stop if the children needed a break, because he would probably have to rush to another class; a breakdown of projector or film loop could upset his whole schedule. By contrast, if the class teacher carried out the testing, she could adjust the rate of questioning to the children, could stop when the children were tired or if the projector broke down, and continue later. It would mean that the test was not given in exactly the same way to all classes, which seemed strange to teachers only used to administering norm-referenced tests. Identical testing conditions, the same time for answering questions, the same instructions, etc., are thought to be necessary for a test to be fair. But equality in these conditions need not mean equality in opportunity to show the degree of achievement of an objective. Finding out whether a child is capable of solving certain problems in a certain time under certain conditions is a different thing from finding out whether he is capable of solving the problems at all. The latter was the point of the tests used in the trials, so flexibility in rate of administration was encouraged. The same argument applied to the instructions and posing of questions in the commentary to the film. A suggested commentary was provided, but teachers could elaborate or repeat parts of it according to what was needed to help their children understand the problem they were being asked to solve.

Our experience of pilot testing in Scotland convinced us, as nothing else would have done, that at least someone from the locality should administer the tests. The evaluator tried out the tests in several Scottish classes; the children were mostly tolerant, but one boy was overheard explaining that the reason they could not understand the commentary was 'because she speaks English'.

The summer course provided opportunity for the teachers to become familiar with the testing technique. Between three and five trial teachers from each area attended the course, together with the area representative. At the course it was possible to explain little more than how to use the film loops and commentary, because quite an amount of time was used to explain the purpose of the testing and the whole evaluation, and the teachers' part in it. The latter was thought to be essential to enlist teachers' cooperation and help, for we were asking them to undertake a very great deal of work, without which the evaluation could not have been carried out. After the summer holiday trial teachers in most areas met at local centres to practice using a loop projector and synchronizing the commentary with the film.

S. 5–13.—2*

The general test and one of the unit tests were given to each child in both the trial group and the control group at the beginning and end of the trial period. The same test was used on both occasions—the limited advantages of using two parallel forms of the test were not considered to justify the extra expense and time involved in producing them. In most cases the tests were administered by the class teacher, who was supplied with a suggested film commentary as well as the film loops and children's booklets. The teacher also marked the tests, using the mark scheme and score sheets provided. As the marking was objective throughout, re-marking by the project team was not thought necessary.

The activities which were going on in the trial areas relating to the evaluation of the first set of units are summarized in Fig. 10 on page 54. For three of the units, the trials lasted for about one and a half terms—from the middle of the autumn term to the end of the spring term. For the fourth unit, *Trees*, the trials continued into the summer term to give the children a chance to see the trees in different seasons. The programme for the evaluation is summarized below in Fig. 5.

	September 1969 to October 1969	October 1969 to March 1970 (June for *Trees*)	March 1970 to April 1970 (June–July for *Trees*)
Trial classes	All took both unit and general tests	Children's science activities guided by teacher using trial unit	All took both unit and general tests
Control classes		Children's activities continued as previously	

Fig. 5   Summary of programme for evaluation of first units

Local education authorities made available a sum of money of about £10 per class to help with extra costs of materials which might be needed for the trial work. This practice continued throughout all the later trials of Science 5–13 units. The way in which the money was distributed varied; some LEAs allowed teachers to use it directly, others made it available to the teachers' centre which bought material in bulk which could then be taken away without charge. One LEA thoughtfully supplied the £10 to control as well as to trial schools, thus controlling what may well have been quite an effective variable in the experiment.

Results of evaluating the trials

The trials were completed as planned, with a remarkably low rate of dropout, as mentioned above. Many problems were encountered at the beginning of the trials. The publication of the trial units was delayed to the end of September but, in the meantime, teachers were pre-testing their classes. The test administration took longer than was anticipated and was far more of a burden for everyone concerned than was intended. The main reason for this was that many of the film loops did not run smoothly; it seemed the films had been inadequately lubricated before being put into the cassettes. Each area had one set of loops for the unit test and one for the general test, which were passed round from school to school with the projector. In most cases films were repaired by someone in the area, which caused less delay to the testing programme than sending for the spare cassettes kept at the project headquarters. The number of incidents of films jamming in cassettes and breakdowns of projectors was greater than had been foreseen; it tried the patience of many teachers, as well as that of the evaluator, almost to the limit. On Form B teachers were given opportunity to report on the testing and a few expressed their feelings clearly:

Whilst the method testing might appear very attractive in theory, in practice I found it to be unreliable, time consuming, and a great tax on one's patience. The children seemed to enjoy the films and generally found the method of answering easy to understand. Nevertheless ... they found the constant interruption very tiresome.

However, the problems were overcome and it was encouraging to find that, despite the trouble, teachers saw the point of it all:

I cannot see how this administration could be made easier, but a section in the unit on time explaining how twenty-four hours in a day could be extended to about thirty hours would have been useful!! To be serious, it meant extra work but we felt it worthwhile.

Between the pre-testing and the post-testing opportunity was taken to make repairs on the films and projectors. The technical problems were much less for the post-testing, which was carried out more quickly and less painfully than the pre-testing.

The numbers of classes remaining in the sample at the end of the trials (with the number at the beginning in brackets) and data about children who completed all the tests on both occasions are given in Table 2. The names of schools taking part are given in Appendix D.

While marking the children's test the teachers made a record on the score sheets provided of each child's performance on every item of the tests. When returned to the project these sheets were used to calculate mean (average) scores for the whole class on the complete test and on separate groups of

Table 2 Numbers of classes and children completing the evaluation trials

Unit	Number of classes				Average age March 1970				Number of children completing all tests	
	Trial		Control		Trial		Control		Trial	Control
					yrs	mths	yrs	mths		
Metals	18	(18)	18	(18)	9	8	9	8	533	502
Working with Wood	17	(17)	17	(17)	10	2	9	11	415	369
Time	25	(26)	25	(26)	9	9	9	11	723	670
Trees	16	(18)	16	(18)	10	0	10	0	438	405
TOTALS	76	(79)	76	(79)					2109	1946

The total number of children taking part in evaluation was 4055

items related to individual objectives of the unit. Class mean scores rather than scores of individual children were the basis for statistical analysis of the test results; the sample size for each unit was thus quite small, being the number of classes not the number of children involved. Means were derived from scores of only those children who completed all of each test on both occasions. In Figs. 6, 7 and 8, the graphs show mean scores for pre- and post-tests for the trial and control groups, for the unit tests.

The graph in Fig. 6 shows that the control classes' scores increased quite considerably, as had been expected; without the control results for comparison it would not have been possible to tell whether the increases in the trial classes' scores were any different from what could have been expected without the trial work. Statistical tests showed that for three of the units, *Working with Wood*, *Metals* and *Trees*, the difference between post-test scores for the trial and control group was significant at the 0.05 level. What this means is that the difference is large enough for the probability of it having resulted by chance or measurement error is less than five times out of a hundred. Difference between pre-test scores for these units were not statistically significant. The results for the fourth unit, *Time*, were complicated by considerable initial differences between the trial and control groups, but they showed the same trend of greater rate of increase for the trial group than for the control. Further statistical tests (analyses of variance) showed that, for the pre-test scores on the unit test, the variation which could be attributed to differences between trial and control were negligible for the pre-test, but was highly significant (beyond the 0.001 level) for the post-test results.

The overall test results were satisfying for the project, since they indicated that the materials were helping children achieve the objectives. It was not a

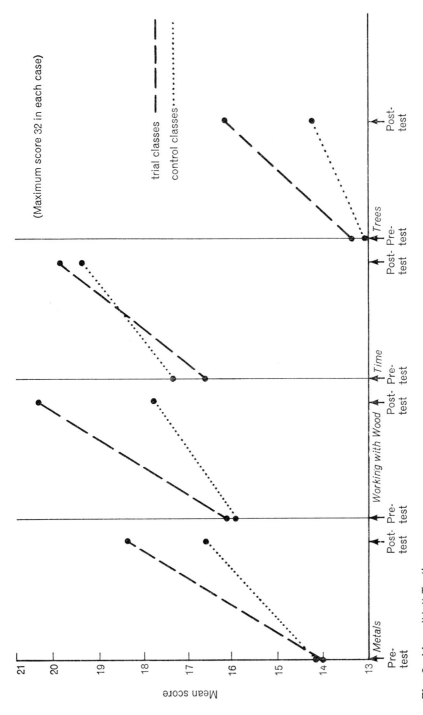

Fig. 6 Mean 'Unit Test' scores

foregone conclusion that the trial classes would improve more than the control classes because—as has been explained—the tests were related to achievement of the objectives, not to learning facts from particular activities. However, these results did not really serve the essential purpose of the evaluation, which was to help improve the units. It was necessary to look not only at overall scores but at sub-scores on items relating to the achievement of individual objectives, so that it would be possible to see whether certain objectives were being more readily achieved than others. Statistical analyses carried out on these sub-scores included a two-way analysis of variance to partition variance between the differences between objectives and differences between trial and control groups.

These analyses revealed considerable variation in the performance of the two groups from one objective to another. For instance, in the case of *Working with Wood*, the change in sub-score for the objective 'awareness of the organized structure of wood' was about the same for the trial and control groups but, for the objective 'awareness of the causes of changes occurring in wood', there was a considerably greater improvement for the trial than the control group. By comparing the changes in sub-scores on the individual objectives, guidance was given as to the relative effectiveness of the parts of the units relating to these objectives. But this evidence could only point to sections that were less effective than others; it could not suggest improvements. For indications of desirable changes other data had to be examined. But before turning to the information from report forms and questionnaires we look quickly at what was found from the General Test results.

Figure 7 shows the overall mean scores for the first part of the General Test. There is a larger scale on the vertical axis than in Fig. 6, and all the differences in score are much smaller than for the Unit Tests. None of the differences between trial and control for pre- and post-tests is statistically significant. For each unit group the General Test scores were broken down into sub-scores on items relating to individual objectives. The results gave indications of whether progress was or was not made towards achievement of the general objectives, suggesting aspects of the activities which might well be given attention in revising the units.

In the second part of the General Test the children expressed their liking for various activities, some involving science and others not. The responses were scored by giving three points for 'liking very much', two for 'liking a little' and one for 'not liking'. Figure 8 gives the mean scores for the nine activities which had most to do with science.

All the mean scores dropped from pre- to post-test, a finding not uncommon in attitude testing. Although the average of the class mean scores decreased, there were in both groups classes whose mean scores increased during the trial period. When numbers of classes with an increase in score were examined for each separate unit sample there was no statistical difference between trial

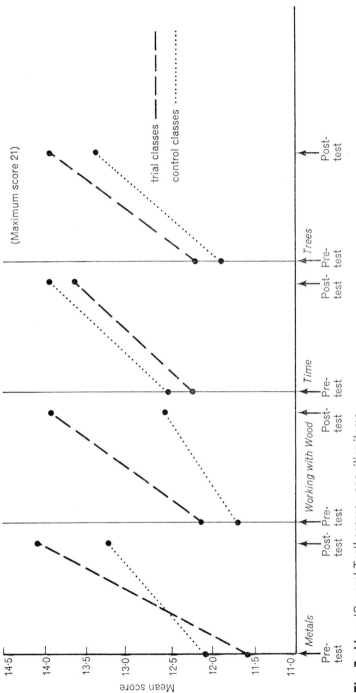

Fig. 7 Mean 'General Test' scores—cognitive items.

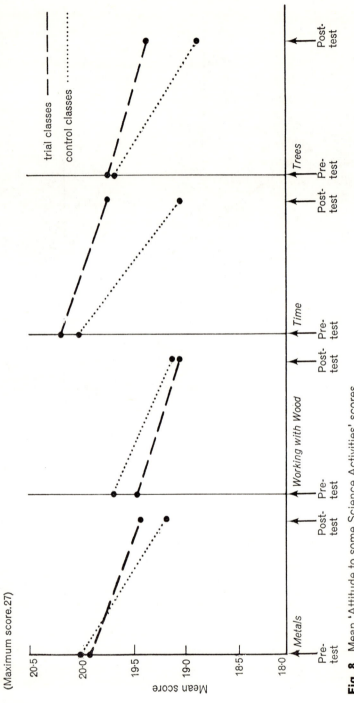

Fig. 8 Mean 'Attitude to some Science Activities' scores

and control groups. When results for the total sample of classes involved in the trials were taken together, however, there was a significant difference in favour of the trial classes. Over the whole sample there was a tendency for more trial classes than control classes to show an increase in mean score on these items, indicating a positive effect of the trial work in increasing children's liking for activities involved in science.

The three report forms filled in by the teachers and the visit report form completed by the team members contained two kinds of information. One was information which was pre-coded, or responses which could be easily coded, and one was information which could not be coded, being either accounts of work done or suggestions about particular activities. After the forms had all been read thoroughly to give a picture of the various types of information they contained the items which could be coded were separated from the rest. The information given on Form A, the diary of activities during the trial period, was passed directly to the authors of the units, and the comments, opinions and suggestions were collected together on file cards. The remaining information was coded, so that the data for each class and teacher was translated into a series of numbers, and then punched on to computer cards. Possible items for each class were numbered, up to about three hundred, and these were analysed by the university computer using a programme for classifying qualitative data. The method of analysis was devised by M. A. Brimer, head of the University School of Education Research Unit, and programmed by R. H. Thomason and I. Parsons of the Computer Unit.

The use of this computer programme was a key feature in the analysis of evaluation information for all the Science 5–13 trials. It revealed patterns of responses or conditions, and enabled the identification of sets of circumstances which were associated with the use of the units in certain ways. The analysis was particularly helpful in the treatment of questionnaire responses which is otherwise very clumsy, since on almost every question there is a range of opinion, and mere totals of replies in any one category convey little more than that some like one thing and some like another. It is much more helpful to know whether a certain reply to one question tends to occur with a certain reply to another, to know whether there are patterns in the responses and, if so, what kinds of conditions are associated with particular patterns. The programme used set the computer to search through the lists of code numbers for each class to find numbers which occurred together in the lists more often than would be expected to happen by chance. Many groups of numbers with a higher coincidence than expected were generally found, and several of these overlapped one another. A second part of the programme made it possible to pick out easily the two groups which represented the ends of the chief dimension or theme running through the data.

An idea of the kind of results obtained from using this programme may be found by taking an example. Some of the items in the groups at the ends of the chief dimension would relate to general classroom working arrangements. These items were among those grouped at one end:

The desks or tables in the room were arranged in irregular groups.
The timetable was fully integrated.
The children worked individually or in groups at most times of the day at their own tasks.
The children formed their own groups for working
The children were able to work on their own ploys.

These items were among those grouped at the other end of the dimension:

The desks or tables in the room were arranged in regular groups.
The children worked as a whole class at most times.
The children's activities were largely directed by the teacher.
The children were allocated to groups by the teacher.
The children all worked on very much the same problem as each other.

Most people would expect the items in these groups to go together; it would have been surprising if, for instance, 'the children were able to work on their own ploys' had been in the second group rather than the first. Nevertheless, it is satisfying when the intuitive feeling that these things go together is confirmed by an objective classification technique. This is all that grouping involved—sorting out the information to find the things that went along with each other. There is no cause and effect relationship between items in the groups, of course; the grouping is merely a statement that the items were found to go together.

While one could have guessed at the way the information about working arrangements in this example would have been grouped, there was very little basis on which to judge how opinions about the project and its ideas would be grouped. Would an item such as 'the teacher made use of the "Statement of Objectives for Children Learning Science"' cluster with the first or the second group? What opinions of the project's ideas would go along with this? Would certain ideas about or reactions to the units be found to be associated together? These were the questions to which answers were needed in order to evaluate the materials and revise them. Information about the arrangements for working and other conditions of the learning environment was included to find out which opinions and reactions tended to be associated, if at all, and with which conditions.

The classification analysis programme was used separately with five different sets of information: the sets of all information relating to each of the four units, and the set of information which concerned working conditions and general attitude to the project not specific to any unit. In all cases the chief dimension discovered was a similar one. It had at one end items which described the material being used effectively and the teacher being satisfied with it, these things going together with children learning through discovery, taking some responsibility for their learning and being able to use their own ideas. At the other end the items described dissatisfaction with the unit, which was not used effectively, grouped together with indications that the children's work was strongly directed by the teacher.

Finding close resemblance between dimensions derived from separate samples of classes increased the confidence which could be placed in the results from any one of them. Similarity between the results of the combined sample and the separate unit analyses was to be expected because of the overlap of the data being analysed, but between analyses for individual units there was no such overlap. So, while the results from each analysis on its own may have been unreliable, the agreement with the three other independent analyses allowed us to use these results with greater confidence.

The combined sample results provided the answers to the questions about how opinions about the project's ideas and objectives lined up with the use of discovery oriented methods. The data for this analysis came from teachers' Form B, from the team members' report form, and from the change in children's test scores. At one end of the chief dimension these items were grouped together:

The children's attitude to science activities test score increased during the trial.

The class had previously been working through active discovery methods.

The desks or tables in the classroom were arranged in irregular groups.

The class timetable was fully integrated.

At most times of the day the children worked individually or in groups at their own tasks.

The children worked regularly outside the classroom.

The children formed their own groups for working.

Science activities were carried on by different groups at different times as chosen by the children.

The children could work on their own ploys.

The teacher had warmly approved the project's ideas on first being introduced to them.

The teacher appreciated very well the meaning of objectives.

The teacher used a discovery approach in most areas of the curriculum.

The teacher had made some use of the project's 'Statement of Objectives' apart from in connection with the trial work.

At the other end were the following.

The class had been previously used to working through formal methods.

The children's activities were very largely directed by the teacher.

The desks or tables in the classroom were arranged in regular groups or rows.

The children worked as a whole class at most times of the day.

Science activities were organized so that all groups always worked at the same time.

Science activities were organized so that all the children worked on much the same problem as each other.

The teacher allocated the children to groups for science activities.

Science activities had not been included in the children's work before the trials.

The teacher had a poor appreciation of what objectives are.

The teacher used a discovery approach hardly at all.

The teacher thought that evaluation of whether or not objectives have been achieved was unnecessary.

The analyses for each unit used information from the teachers' Form B and Form C, D, E or F, according to the unit concerned, as well as information from the team member's report and from the change in children's test scores. The results made it possible to see which opinions and comments about the unit tended to go together with general satisfaction or dissatisfaction. This could be done for every item of information, not just those which were found to be associated in a group, since the computer programme provided a numerical weighting which indicated the degree of association between any item in the data with the items in a group. Thus, we could turn to any item of the questionnaire and find whether any of the responses were associated with groups at one or other end of the dimension. For example, in the report form on *Working with Wood*, this question was asked:

In the section on activities would you have preferred:

(a) fewer suggestions in more detail?
(b) more suggestions in less detail?
(c) no change?

(Note: This item followed others in which the questions of number and detail of activities had been investigated separately.) It was found that response (a) was negatively associated with the satisfied end of the dimension and was in fact included in the group at the dissatisfied end. Response (b) was positively and quite strongly associated with the satisfied end, and negatively with the dissatisfied end. Response (c) had a low negative association with both ends.

It was also possible to estimate from the weighting of responses for individual teachers and classes where each could be placed along the chief dimension. In reading a teacher's comment, suggestion or criticism, we therefore knew to what extent the teacher making it was satisfied and had made good use of the unit, or was dissatisfied and had not been helped much by the unit. The comments had much greater value when examined in this way than if they had all been treated as coming from the same background of use and opinion of the project's materials. Obviously a comment, either adverse or favourable, would be received differently according to whether it came from someone who had tried to use the materials as intended or someone who had used new materials in old ways.

Making use of the results

The procedure for communicating results of the evaluation to the unit authors remained much the same for all four sets of trials. The evaluator wrote a confidential report on the results for each unit which was circulated strictly among team members only. The report gave a list of the items in the groups at either end of the chief dimension and then the degree of association of other items with these two groups. Comments from teachers were used to illuminate the meaning of many of the items. For example, an item might be that 'Some activities were thought to be too difficult' or 'too easy'. This was not very helpful unless it was also known which activities were being criticized. Such information was contained in the free responses given on the forms and these were retrieved from the file cards in which they had been collected together. In the case of the first set of units, the report also contained results of analysis of children's test scores and an interpretation of them in terms of pointers for parts of the unit which should be given attention in rewriting.

It has to be admitted that the value of the test results in revising the unit was not very great. The testing had been useful in indicating overall achievement of stated objectives and had thus served a useful purpose. As a guide to making revisions to the units, however, the test results could only point to certain sections of the units where there had been little progress toward achieving the objective, but they could not suggest why there had been no progress or indicate ways in which these sections might be changed. In addition, it was difficult to relate change in the test scores of a particular class to the effect of experiences during the trial period. Generally it was found that children already used to discovery learning would score more highly on the pre-test than children of similar social background and age who previously had been used to teacher-directed learning. If both reached the same score level on post test, one class would show a greater change in score than the other, but this may be a reflection of previous experiences more than of the effect of any experiences during the trial period.

By far the most help for revising units came from the analysis of teachers' report forms and the team members' report form. Being able to identify groups of responses or items of information occurring together brought an order to these data which might otherwise have remained a disorganized mass. After finding the association between each item of opinion or comment and the two groups at the ends of the chief dimension, we then knew which ones were linked with dissatisfaction with the unit. These responses were treated as reasons for dissatisfaction, and they immediately indicated points which required attention in rewriting and suggested the kind of change that should be made. For instance, the item 'The teacher wanted help (with class management and group organization) which was not supplied by the unit' was in the dissatisfied group or strongly associated with it for all units. The cause of dissatisfaction could be removed, in this case, by writing in suggestions which provided the kind of help which was felt lacking. Usually it was possible to find ways of correcting causes of dissatisfaction without upsetting those who were already satisfied; in this example it is unlikely that including help with class orgainzation would upset those who had been able to make good use of the units without it.

To go into detail of revisions made in all the units, and the evidence on which decisions to make the changes were based, is not appropriate here where we are attempting only to describe the procedures. However, a summary of changes made in one unit may help to illustrate the process of rewriting; the unit *Working with Wood* provides a good example.

1 The material in this unit was completely reorganized in response to the finding that teachers felt the topic was too narrow to suit the wide ranging exploration which is appropriate in the junior years. The unit author, using her own first hand observations in the trial classes as well as the evaluation findings, rewrote the activities so that they would fit into a wider context and she indicated how they could lead to and arise from other classroom activities.

2 More help was given with the details of organizing activities in the classroom and dealing with some problems which might arise in practice.

3 In reorganizing the activities, the separation into Stage 1 and Stage 2 activities was broken down. Teachers not satisfied with the unit had found this separation was not helpful because, in order to cater for children at different stages, they had to be constantly referring to different parts of the book and to different activities. It was felt more helpful to point out, in rewriting, how one activity could be used to achieve various objectives at both Stages 1 and 2.

4 The arrangement of activities grouped under the heading of one of the objectives appeared to have been a source of dissatisfaction. Teachers frequently pointed out in their comments that keeping one objective in mind seemed to mean that children missed chances of progress towards achieving other objectives. This was, of course, the reverse of what was intended, so different means of indicating objectives were sought.

5 More help with ways of recording work was requested by teachers. Recording was something which most teachers reported their children

did not usually like to do. It was clear from many accounts that, during the trials, there had been too much emphasis on conventional written records. In revising the unit, suggestions were made—often in the form of illustrations of children's work—of a variety of ways of recording which would be more attractive and equally valuable to the children.

6   Many of the activities in the unit were modified or extended as a result of suggestions or comments from the trial teachers. The few unsatisfactory activities were taken out; new activities were added, often ones which had been developed by teachers as a result of extending the activities during the trials.

Final decisions as to whether changes recommended by the evaluation should be made rested with the team in general and the unit author in particular. Generally the indications that the author had picked up during visits to the trial classes supported the evaluation findings and were supported by them. Where this was not so the author had to decide whether the evaluation made out a strong enough case for making certain changes. The evaluator's job was to provide information and to suggest decisions based on this information, but the final decision making had to be left with writers.

After the results had been considered and used, a report was written specifically for the teachers and administrators who took part in, or helped to organize, the trials. There was not time available to write a full technical report—we were too busy making use of the results and looking after subsequent trials—but it was important to make time to feed back information to those who had made the evaluation possible. A short account was therefore given in a 'Report for Teachers on the Evaluation of the First Set of Units' (1971), and copies were sent to everyone who had taken part, whether as an administrator or as a trial or control class teacher.

4 Evaluating trials of the second set of units

Planning the evaluation

Writing the second set of units began when most of the units in the first set had reached the stage of pilot trials. Production of the two sets thus over-lapped, as did their trials, since trials of the second set began when those of the first set were in progress. Planning the evaluation of the second set took place at about the time the first set trials began, and thus it was not possible to apply experience gained from the earlier trials to the procedures adopted for the second ones. The theoretical arguments which led to decisions about what to evaluate for the first set of units therefore remained unchanged and were applied equally to the units of the second set. However, there were practical considerations which restricted the scope of the evaluation in the case of the two Stage 1 and 2 units of the second set.

The argument that changes in children's behaviour should be measured because the purpose of the material was to promote achievement of behavioural objectives was still felt to be a strong one. Equally strongly it was held that conventional paper and pencil tests would not give valid measures of the kinds of behaviour under investigation and would have considerable disadvantages in terms of affecting adversely both the children's and teachers' enjoyment of the work. This problem has already been mentioned on page 27, where it was concluded that the use of film loops to introduce test items would meet many of the objections to tests.

However, it became evident at the summer course preceding the first set trials that the practical problems of obtaining a suitable projector and arrang-ing its transport from school to school, though not insuperable, would be quite a burden for the area representative. Then, when testing began, the ensuing chain of various technical breakdowns made it evident that the testing was a greater problem than had been anticipated. We could not consider doubling the difficuties by introducing tests in another six schools simultane-ously with first trials. So, rather than resort to what was thought would be a less valid type of test, it was decided to omit children's tests from the plan for evaluating Stage 1 and 2 units of the second set.

It was not felt that too much would be lost by this decision. After all, these two Stage 1 and 2 units were not the first to be produced or given trials. The four Stage 1 and 2 units of the first set were being evaluated and it was expected that a considerable amount of the information obtained would be relevant not only to these four but to other Stage 1 and 2 units as well. In the event it turned out that, fortuitously, we carried out a test of the value of results from children's tests for the purposes of the evaluation. Two sets of trials were conducted, one with the testing and one without it. At the end of the last chapter it was concluded that the testing gave little positive help to guide rewriting; the trials of the second set confirmed this by showing that the other information gathered was by itself quite adequate for the revision of the units.

The same argument did not, however, apply to the third unit of the second set, a unit written specifically for infants' school teachers and entitled *Early Experiences*. This unit had been begun after pilot trials of the first set Stage 1 and 2 units had shown that these were not as helpful to infants' teachers as had been hoped. Work too closely tied to one theme was not appropriate; infants' interests jump quickly and range widely from one thing to another, and science is so much interwoven into activities that it is not distinguishable as a separate kind of study. *Early Experiences* was an attempt to guide teachers to encourage scientific thinking and achievement of 'ground level' objectives in science while using teaching methods and materials appropriate to infants. It was the only unit of its kind, so its unique position meant that findings from other trials might not necessarily be readily applied to it.

It was felt that, in the case of *Early Experiences*, an attempt should be made to see if there were signs of progress towards achievement of the unit's objectives on the part of children taking part in the trials. At the same time it was appreciated that obtaining reliable measures of behaviour changes in young children is very difficult and that any results would be clouded by a large number of uncertainties. Individual testing was necessary with the very young children involved and, because manpower for administering tests was very limited, it was decided to test only a small sample of children from two trial areas. The testing is described briefly on p. 50.

The results from the testing were not expected to be very decisive, since changes in children's concepts, attitudes and abilities are not made overnight. We were particularly aware that, because the unit *Early Experiences* tried to draw out scientific aspects of normal activities and did not suggest a course of 'science activities', it might be necessary for teachers to have been using the unit for some extended time before changes were likely to be apparent in their pupils. On the other hand, changes would not occur in the children unless teachers were able to use the unit in a way which was expected to promote achievement of the objectives. Therefore one measure of the effectiveness of the unit would be the degree to which it increased teachers' willingness to provide conditions and opportunities for children to have experiences which would help them achieve the objectives in the long term. An instrument called the teachers' 'preferences form' was developed to investigate this effect upon teachers taking part in the trials.

The evaluation materials

For the two Stage 1 and 2 units, *Structures and Forces* and *Science from Toys*, the evaluation materials consisted of questionnaires and report forms very similar to those used for the first set of units:

Form A: mainly blank pages for a diary of relevant events in the trial class during the trials.

Form B: a questionnaire, mostly pre-coded, asking for information about the class and school, relevant biographical data about the teacher, and opinions of the book *With Objectives in Mind*.

Form G or H:

questionnaires, mostly pre-coded but with provision for free responses as well, asking for information, opinions and suggestions about the unit. Only one of these two was completed according to which unit was being tried.

Team members' form:

completed by team member, after talking to the teacher and visiting the class when trial work was in progress.

In the case of *Early Experiences*, a new aspect—the change in the teachers as indicated by the Preferences Form—had been introduced. The evaluation materials and their purposes are indicated in Fig. 9.

It is convenient here to use the word 'test', but in the trials we avoided the word and instead called the situations in which the children's abilities were judged 'selected activities'. Each selected activity was a practical situation posing a problem for the child which he was asked to solve by doing something, manipulating the given objects or picking out certain things from others. From the way the child acted and what he said about it, the person presenting the activities judged his ability to solve a problem or his grasp of certain ideas.

The selected activities were administered by the evaluator and another team member to a randomly selected sample of children from the trial areas, Bristol and Kent. The same items were given at the beginning of the trial period and again to the same children at the end. The same tests were given to an equal number of children randomly selected from infants' school classes in the same areas but not taking part in the trials. These acted as controls so that it was possible to estimate how much of any changes in performance could have been the result of factors other than the trial work. The items of 'selected activities' covered the cognitive objectives in the main, though one item was designed to find out about children's liking for activities involving

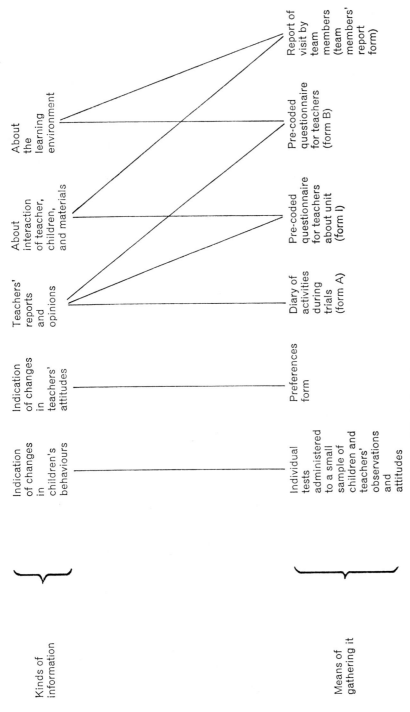

Fig. 9 Information gathered in the evaluation of *Early Experiences*

active exploration. But by being with the children throughout the trial period, the teacher was in a better position than the visiting team member to judge changes in children's attitudes. A form was therefore provided for teachers to record children's willingness to do certain things, to question and to investigate, at the beginning and again at the end of the trials. These were filled in only for the sample of children tested in the trial and control classes.

The teachers' Preferences Form was developed after a survey of existing measures of teacher attitude had shown that there was none suitable for indicating attitudes towards science activities for young children. It was also felt that conventional attitude scales would be unsuitable because they might arouse suspicion or cause anxiety, either of which could encourage 'faking'. Since attitudes can be inferred from behaviour, a decision was taken to develop an instrument in which teachers would make evident their attitudes by expressing their preferences for certain ways of working with children. All the choices were made 'positive', that is, concerned with taking action rather than avoiding action, and were expressed so that they were all acceptable in some circumstances. The selection made from these choices would, it was argued, give a true account of the teachers' preferences, since there was no reason to falsify results nor any obvious selection of choices indicating a supposedly favourable 'image'.

As a result of changes made after small scale trials of early drafts, this instrument took the form of statements divided into three lists, each dealing with a main aspect of the classroom situation: working arrangements for science activities, children's activities, and the provision and use of materials and equipment. Each contained twenty statements and respondents were instructed to tick the ten from each list which most correspond with the things they do or prefer to do with their children. After several revisions, the form was given to one hundred and twenty-eight primary school teachers for the purpose of developing a 'score' or weighting for each item. The computer programme for classifying qualitative data, described on page 41, was used. The chief dimension which emerged very distinctly from the analysis related to the freedom and responsibility for learning allowed to individual children by their teachers.

The weighting derived for each item in the preferences lists indicated its association with a child centred approach to science activities. At the low end of the scale were items which indicated restricting the children, both physically, to work at their desks, and mentally, by directing activities rather than allowing choice, by preferring inactive learning situations to active ones. At this end were preferences for obtaining 'correct' results and making accurate written records. The opposite approach was found at the high end of the scale: allowing children some freedom of choice for activities, being flexible about when and where science activities took place, encouraging discussion as an alternative to written communication, giving the children responsibility for their work and for equipment and confidence in their observations. The Preferences Form measures were validated in separate trials in which the same sixty-eight teachers completed the Preferences Form and the 'Survey of Opinions about Education' form produced by R. A. C. Oliver.[1] Scores on three scales are given by Oliver's instrument: tendermindedness, radicalism and naturalism. Correlations between Preferences scores and all three of these scales were positive and statistically significant (at the 0.01 level). The highest correlation was with naturalism, closely followed by tendermindedness.

Arrangements for trials

Proposals for evaluating trials of the second set of units were circulated to trial area representatives and discussed at a meeting in the term before the trials began. Representatives were given the choice of trying *Science from Toys* or *Structures and Forces* in up to six junior classes, while *Early Experiences* was tried in up to four infants' classes in every area. Criteria for choosing classes were outlined as before so that the sample would encompass a spread of age levels and types of school. Some areas preferred to keep the number of classes to less than the maximum suggested because of the pressure of the first trials which were already in progress. Seven areas chose to try *Structures and Forces* and five *Science from Toys*; Scotland tried only *Early Experiences*. Selection of classes in most areas was simplified since no control classes were involved, and choosing the control classes in Bristol and Kent where the 'selected activities' were to be administered did not present any problems.

In the holiday before the second set trials began in January 1970 a one-week course was held for teachers taking part in the trials. As in all cases, there were not enough places on the course for all trial teachers from all trial areas to attend. However, it was by then the established pattern that the area representatives and those teachers who had attended would, on returning home, organize meetings and workshops for all trial teachers to become equally familiar with the units and evaluation materials. In many areas non-trial teachers were also welcomed at these local meetings and used the units, although not taking part in the evaluation trials. Thus, dissemination was begun at an early stage in the project's life while the material was still at the draft stage.

Figure 10 indicates the main events going on in the trial areas in the school year 1969–70 when the first two sets of trials were in progress. It gives only an outline of the activity during this time and does not include either the meetings between teachers and area representatives which were going on throughout the year, or the visits of team members to the areas. The programme for the evaluation of the units of the second set is summarized in Figs. 11 and 12.

	Courses and meetings with trial teachers	Six junior trial classes	Six junior control classes	Six junior trial classes	Four infants' classes
1969 Aug					
	Course for trial teachers				
Sept		Pre-test units, general	Pre-test units, general		
Oct		T R I A L			
Nov					
Dec	Meeting of area representatives	W O R K			
1970 Jan		I N		T R I A L	Pre-tests in two areas
Feb	Course for trial teachers	P R O G R E S S		W O R K	T R I A L
Mar					W O R K
Apr		Post-tests for 3 units	Post-tests for 3 units	I N	
May	Science 5–13 course			P R O G R E S S	I N
June					P R O G R E S S
July	Meeting of area representatives	Post-test for *Trees*	Post-test for *Trees*		Post-tests in two areas

Fig. 10 Trial area activities 1969–70

	Early January 1970	January to June 1970	July 1970
Trial classes in all areas (No controls)		Teachers use either *Structures and Forces* or *Science from Toys*	Forms A, B, G or H completed
		Team members' visit and complete report forms	

Fig. 11   Programme for trials of the second set of Stage 1 and 2 units

	All trial areas	Bristol and Kent	
		Trial	Control
Teachers use *Early Experiences* January to June 1970	√	√	
Form A completed June/July 1970	√	√	
Form B completed June/July 1970	√	√	
Form I completed June/July 1970	√	√	
Preferences form completed January 1970 July 1970	√ √	√ √	√ √
Selected activities given to sample of children: January 1970 July 1970		√ √	√ √
Teacher's record attitude of children tested: January 1970 July 1970		√ √	√ √
Team members visit and complete report form	√	√	

Fig. 12   Programme for trials of *Early Experiences*

Results from the trials

For the two Stage 1 and 2 units the results were all from questionnaires and report forms, and they were treated in exactly the same way as has been described for the units of the first set (see pages 41 to 43). Separate analyses were carried out for each unit and the chief dimensions identified were again similar to those found in analyses for the first set of units. For instance, in the case of *Structures and Forces* some of the thirteen items grouped together at one end of the dimension were:

The teacher grasped the idea of objectives fairly well.
The teacher felt she could not have made use of the unit quite as easily without *With Objectives in Mind*.
The teacher found the statement of 'Objectives for Children learning Science' very useful.
The teacher thought consideration of educational aims important.
None of the activities in the unit was avoided because considered unsuitable.

Some of the twenty-three items grouped at the other end were:

The teacher grasped the idea of objectives rather poorly.
The class had previously been used to formal teaching methods.
The teacher used a discovery method hardly at all.
The tables and desks in the room were arranged in rows or in a regular array.
The unit was considered too brief to serve its purpose efficiently.
The teacher thought the indications in the unit of how the work could lead to activities in other areas of the curriculum were inadequate.

All the responses in report Forms G and H were examined for their association with the groups at either end of the chief dimension. As before, those responses associated with dissatisfaction were regarded as possible reasons for some teachers not finding the unit helpful and attention was focused on these points in rewriting. Suggestions and criticisms from teachers were particularly helpful in pointing out ways of improving parts of the unit. The computer anaysis results enabled the teachers' responses to be summarized by finding their position along the chief dimension; this information was the greatest help in interpreting opinions and suggestions made as free responses in the questionnaires. As before, the results and possible interpretations were written into a confidential report for team members only.

In addition to the information provided by the analysis there were, of course, also the subjective impressions the authors had gathered from seeing work in progress during the trials. Generally one kind of evidence helped to support the other; the analysis of responses would help to strengthen or explain certain 'feelings' the author had about various parts of the unit. In few cases was there conflict between the two kinds of evidence, but one instance of this led to important changes in the unit *Science from Toys*.

Some teachers reported that sections of the book, including the one on 'Boats', were too difficult; yet it seemed that the children had done some good work and of the kind expected. However, a closer look at the teachers' comments revealed a misunderstanding. Parts of the text which had been intended as background information for the teacher were read as teaching material. Hence the teachers were critical of the book and dissatisfied with the children's work. In rewriting this unit, one of the changes made was to clarify the distinction between 'background' and 'classroom' material.

Turning now to the results for *Early Experiences*, it should be said straight away that—rather as expected—administration of the 'selected activities' gave almost no information that was relevant to rewriting the unit. Children's scores showed that there was very little difference in overall score between the sample of thirty trial and thirty control children. Sub-scores for each objective included in the test revealed differences between the groups, but none large enought to be statistically significant. There was most change in performance of trial relative to control children in the case of the objectives: 'awareness of human characteristics', 'ability to discriminate between different materials', and 'ability to record events in their sequence'. (The objectives were revised after the trials of this unit; some of these objectives are not the same as the ones in the published version of the unit.)

The teachers' estimate of changes in attitude among the children could not be analysed statistically, since only eight trial and eight control teachers were involved, and the judgements must have depended very much on the standards applied by individual teachers. There was negligible difference between the total number of changes of a positive kind reported by the trial and control teachers. On individual attitude objectives, there was a greater improvement of trial over control only in the case of 'willingness to explore non-living things'. These findings were highly inconclusive; investigation on a larger scale and over a longer period was required to decide whether intended behaviour changes were appearing in the children.

The remaining evidence, from the questionnaires and report forms, was thus used for guiding the revision of the unit. It was analysed as has been described for the other units, the only difference being the inclusion of data from the teachers' Preferences Forms. The Preferences Form responses were scored by summarizing the weightings of each item ticked (see page 52). The resulting scores for the first and second time the form was completed were used to supply two kinds of information: about the change in Preferences score, and about the initial level of score. The change in score was classified, and coded for the computer analysis, as being:

positive if it was greater than $+2\cdot00$
    zero if it was between $-1\cdot99$ and $+1\cdot99$
negative if it was greater than $-2\cdot00$

The total range of initial scores was divided into five bands, each band having a different code number. So, for each class and teacher, the information analysed included responses from Forms B and I and from the team members' form, and two numbers indicating initial level and any change in Preferences score.

The chief dimension emerging in the data for this unit indicated far less variation among infants' teachers in respect of grasping objectives than had been found among teachers of older children. At one end items signifying satisfaction with the unit were grouped with indications that individual learning and discovery work were being managed successfully. At the other end dissatisfaction went together with evidence of less attempt to organize, and less success in managing individual discovery work. High initial Preferences scores were strongly associated with the satisfied end of the dimension while low initial Preferences scores were part of the group at the dissatisfied end. Changes in Preferences scores were not associated strongly with either end.

As for other units, the information about the items in the groups at the ends of the dimension and the association of responses in Form I with these groups was supplied to the author and other team members in an internal report. The position of each teacher along the dimension was found and enabled the author to interpret teachers' opinions against the background in which they were formed. There were in fact few really adverse comments about the activities in this unit, the most likely reason being because it had been through extensive pre-pilot trials, during which many changes had already been made. The main remaining changes found desirable were in the direction of providing more arguments, encouragement and help for discovery methods and in pointing out how the scientific aspects of usual infant activities could form the starting points for the work.

The coincidence of low initial Preferences scores with items indicating dissatisfaction showed how important for using the unit was a teacher's attitude towards allowing children freedom to learn from their own activities. It was realized that attitudes are not easily changed and that, for many teachers to make effective use of the unit, help may be required in a form other than the written word. While as much as possible was done in the unit by including case studies and photographs to illustrate the value of the approach, it was suspected that, for some, the unit on its own would not give enough help. Either an introductory course or some supporting material was needed to develop attitudes which favoured effective use of the unit. The same argument, supported by evidence from later trials, applied equally to the project's other units and, in a subsequent stage of the project, the development of supporting material in the form of the unit *Understanding Science 5–13* was undertaken. (See page 89.)

After the interpretation and use of the results, another report was written for the teachers and others who had helped with the trials. This one was entitled 'Report for Teachers in the Evaluation of the Second Set of Units' (1971) and copies were sent to schools where classes had taken part in the second trials, and also to schools where there were trial classes for the third set of units.

References

1   Oliver, R. A. C., and Butcher, H. J. 'Teachers' attitudes to education: The structure of educational attitudes', *British Journal of Social and Clinical Psychology*, **1**, 1962, 56–59.

5 A change in the approach of the evaluation

A critical look at the evaluation of the first two sets of units

The last two chapters have shown that the evaluation procedures for the first two sets of units were, of necessity, similar to each other, since there was no feedback available from the first before the trials of the second set began. But between July 1970, when these trials ended, and January 1971, when trials of the third set of units began, there was time for reappraisal of the procedures and materials which had been employed in the evaluation.

As well as using the results from the first and second sets to revise the units, it was also possible to use them to revise the evaluation. Which of the various kinds of information were really helpful in revision became quite evident during the interpretation and use of the results. Similarly, it had been quite evident during the trials which kinds of information took most effort to gather, and it had not been necessary to wait until the end of the trials to become aware of the teachers' criticisms about the evaluation. These criticisms were most certainly not restricted to the practical difficulties and amount of work involved; on the contrary they were in most cases aimed at pointing out how the organization or content of the materials might be obstructing the evaluation from achieving its purpose.

So, emerging either as a result of applying the evaluation findings or from teachers' comments, the following appeared as the five main points of criticism:

1 The organization of the trials, with one teacher trying one of the units only throughout, was preventing the units being used flexibly. It had been realized that keeping to one unit for two terms was not the way the materials were intended to be used in their final form, but it was thought the best way to give them a thorough trial. Many teachers pointed out that they would not normally keep to one range of activities throughout such a period of time and comments showed that some were not happy about working in this artificial way. The arrangement also precluded any comparison between units because no two units were used by any one teacher.

2 During the trials it was not known to what extent the results of testing

the children would be useful, but teachers did note the restricted range of objectives covered by the items. One teacher said, 'The greatest changes that I noticed in my children during the trials were in the number and kinds of questions they asked, in their perseverance and concentration on quite involved problems, and how I began to find my difficulty was in getting them to stop work not to start it—but your tests did not give any chance for these changes to show.' These changes were valued more highly by teachers of young children than progress towards achievement of some of the cognitive objectives. In the absence of a way of detecting such changes an important measure of the effect of the units was being neglected.

3   When the children's test results had been analysed and interpreted it became clear that their value for rewriting was very small in proportion to the effort required to collect them. Even if the many practical difficulties had not been encountered, the results of testing could give only very vague pointers as to which parts of the units were relatively more or less effective; they were incapable of providing explanations or indications of how the various parts should be revised.

4   The technique of analysing many different items of information, using the computer programme (page 41) to reveal groups of items which were found to occur together, also gave the weighting of various kinds of data with the groups (pages 44 and 45). These weightings suggested the relative importance of gathering information of different kinds. Whenever results from classroom visiting, from teachers' questionnaires and from other sources were put together and sorted out using the classification programme, it was found that data from making first hand observations in the classroom were highly weighted and this was the most important factor in facilitating thorough use of other kinds of information. In the first and second trials the information from classrooms had been gathered in a somewhat superficial way; even so, it proved most useful. It was clear that attention should be given to gathering it in a more reliable and organized way.

5   The use of the Preferences Form in the trials of *Early Experiences* had provided interesting results which had contributed to the revision of the unit. It also added substantially to the information to be taken into account when considering the background to teachers' comments and opinions. It was felt that the use of this kind of instrument should have been more widespread in the trials.

A new wave of units: fresh questions for evaluation

During the trials of the first and second set of units the team members were beginning to produce another wave of units, which became the third set and comprised four more Stage 1 and 2 units and the first two Stage 3 units. Results from the first trials were delayed, due to trouble with the university computer, and the units of the third set were produced in trial edition before the earlier units were revised. Nevertheless, in writing the new units the team members were able to make use of their subjective impressions of the way the trials were progressing. Changes in writing style and organization of the unit were brought about by absorbing some of the results of the evaluation even before these had been fully analysed.

Some part of the change in style of the new units is accounted for by their subject matter. Of the four Stage 1 and 2 units, three concerned what might be called 'linking themes' which could be pursued through a variety of topics and materials: *Coloured Things, Change, Holes, Gaps and Cavities*. By contrast, *Working with Wood, Metals* and *Trees* were very much centred round the exploration of specific materials. *Minibeasts* was the only Stage 1 and 2 unit of the third set to be linked to a defined area of study, and even this was loosely defined. In the new units there was, therefore, more emphasis on providing a large number of ideas which exemplified an inquiry approach and from which teachers could select and develop those ideas which were appropriate for their particular children and situation. Immediately the question was raised as to whether the change in emphasis was an improvement.

A further difference between the later and earlier units concerned the way in which objectives potentially achievable through an activity were pointed out. In Chapter 3 it was described how teachers found that the division of activities into those for Stage 1 and those for Stage 2 had serious disadvantages. Also the grouping of activities under the chief objective to which they related had been criticized. In later units it was decided to abandon grouping according to stage and objective, but what could replace the help this grouping had given? It was undoubtedly useful for most teachers to know which objectives they should have in mind when guiding certain activities, and to know what were realistic objectives for children at different stages of development. The problem was how to provide this help without at the same time giving the objectives a role which might restrict the range and richness of the activities.

One step which was agreed as being helpful by all team members was to include the list of objectives, revised once more after the trials of the first two sets of units, at the back of each unit. Teachers would then be able to see which objectives were suggested for Stages 1, 2 and 3, and might select for themselves those relevant for particular activities—if they felt capable and had the time. Some authors felt that further help would be needed by many teachers if only in providing criteria for selecting objectives and suggestions as to how having the objectives in mind might influence the teachers' guidance of the work.

In the four Stage 1 and 2 units different approaches and combinations of approaches to providing this help were tried. For three of the units the potential objectives for each activity, at either Stage 1 or 2, had been identified and used to construct an index to objectives. The index, together with an extensive introduction indicating its purpose and possible uses, was given at the back of the book. The entries were linked to the list of objectives by the use of code numbers for each objective. In the unit *Change Stages 1 and 2*, in addition to the index, the code numbers of the relevant objectives were indicated in the margin of the page beside the suggestions for activities. In two units a discussion on working with objectives was included in the first chapter, and one had a further section on 'thinking—through an objective'. The various approaches and their use are indicated in the summary in Fig. 13.

	Stage 1 and 2 units			
	Coloured Things	Mini-beasts	Change	Holes, Gaps and Cavities
Objectives at back of unit	√	√	√	√
Index of objectives in unit	√	√	√	
Objectives indicated in margins			√	
Chapter on working with objectives	√	√		
Section on 'thinking—through an objective'		√		

Fig. 13   Approaches to indicating objectives in the third set of units

Of course the problem raised many questions for the evaluation to tackle. How successful were the attempts to provide guidance as to objectives? Was one approach any more successful than any other? Many of these questions could only be answered if the trials were organized so that comparison was possible between particluar aspects of the units.

Planning the procedures for more effective evaluation

A pattern of trials organization quite different from that used before was proposed for the evaluation of the Stage 1 and 2 units of the third set. It attempted both to meet the criticism of the earlier pattern by providing for the units to be tried in more 'normal' circumstances and to enable comparisons of units to be made by teachers who had used several during the trials. In the new pattern teachers were gathered into groups, generally of four, sometimes in the same school but more often from different schools. Each teacher used all four units and, through meeting together, the members of the group arranged between them to try out all the units. In this pattern the teachers had far more freedom to pick out parts from various units which would interest their pupils and so could use the units in a way which was much more in keeping with the project's intentions. Another advantage was that, since more teachers were using and reporting on all the units, the sample of opinion about each unit was much larger than before.

As will be seen in the next chapter, when given the choice between the new and the old arrangements for trials, a substantial majority of trial areas chose to adopt the new pattern. In practice, this led to a further advantage of the new pattern, which had not been foreseen and was not revealed until after the end of the third trials. The formation of groups of teachers gave them the opportunity to discuss their work, to visit each others' schools for meetings, to break down the invisible barriers which so often divide one school from another and even one teacher from another in the same school. Futhermore, rather than dispersing at the end of the trials, many of the groups increased their membership; others split up to form more groups, with the trial teachers acting as group leaders. These spontaneous arrangements helped to prepare the way for wider dissemination of the project's materials and ideas and also promoted local initiatives in developing materials, which was one of the project's long term aims.

The teachers' criticisms of the children's tests in the first set of trials, coupled with the finding that the results were relatively ineffective in guiding rewriting, led to a search for alternatives to objective testing of the achievement of objectives. On the whole the need for checking upon and recording the children's progress was appreciated by teachers, it was the narrow range of objectives encompassed by the tests which had attracted criticism. So the objective testing was replaced by the use of a checklist of behaviours which were called 'diagnostic statements', about which more is said later. The dual purposes of providing this instrument were seen as helping teachers make good use of the units and enabling them to record their children's progress. It was not expected that the results would be of much help for revising the units, since the same reasons which had made the results of the objective tests of little value would apply even more strongly to the results of teachers' subjective judgements.

The way in which the analyses of information from the earlier trials enabled us to evaluate the usefulness of data from various sources has been mentioned in point four, on page 6. The results lent support to the idea that it was vital to know how the material was being used before useful interpretation of comments upon it could be made. However, the time which team members could spend in the classrooms making observations about the use of the units was very limited. Visits to trial areas absorbed a large proportion of time while the trials were in progress and necessitated a great many hours of travelling. Sometimes, inevitably, not all the visits on the programme provided the required information—teachers might be ill, or school plays being rehearsed—and making another visit caused a great deal of inconvenience. Evidently it would be much easier organizationally for someone in the locality to make the visit. There would be other benefits too, for the local person could be some-one known both to the teacher and children, whose presence in the class would, for this reason, cause less interruption of the work going on. Also, a local person could more easily make a longer visit, and a second if necessary, and so would have more chance of making an accurate report.

The trial area representatives were thus asked to undertake, or to organize, the observations in the classroom. A 'visitors' report form', a much revised edition of the old 'team members' report form', was provided for the obser-vations to be recorded. More is said about this form below. It was possible to introduce the form and discuss its use with the area representatives at the project course held just before the trial began. There were also extensive 'notes for users' at the beginning of the form. The visiting and reporting went very smoothly in the hands of the area representatives; in many cases they exceeded our demands by carefully double checking the information and, in several areas, two people visited each class for the purpose of reporting.

In this way team members were relieved of the work of recording obser-vations, but they still visited the classes so that they could gain their own impressions of the work and talk to individual teachers about any problems. The team members' impressions also provided some validation of the results in the visitors' report forms.

It is interesting to speculate upon why the arrangement for gathering information from inside the classrooms worked so well at this stage of the project's trials, but we did not feel that it could have been considered earlier. There were many advantages of running several sets of trials instead of trying all the materials together, the chief ones being that we could learn from our earlier experience in evaluation and that the organization became smoother and more efficient through repeated use. With few exceptions the trial area repre-sentatives remained the same throughout the trials, there was a very high rate of attendance at meetings with the team, and communications with the project were close at all times. By the time the third trials were undertaken the representatives had had experience of organizing the two earlier trials and had set up local arrangements for courses and meetings. They had been to three project courses and had been absorbing the project's ideas for at least twenty months. With this background they were well able to pick up in a classroom the information which would help judge how well the project's ideas were being applied. In cases where the representative delegated the visiting to someone else in the area it was invariably to a person who had been assisting earlier trials and had attended courses.

Another reason may have been that participation in earlier trials showed that the project was making a great deal of use of the findings. The project genuinely took the trials seriously—not as a rubber stamp for preconceived ideas; the results arising from help given by teachers and administrators in the trial areas were being seen in the changes made in the units, and were being fed back directly in reports on the trials. Possibly, then, the area administrators felt willing to give their help because the project evidently needed it and, equally evidently, would make good use of it.

Evaluation materials used in the third set of trials

THE TEACHERS' REPORT FORM

Analysis of responses to teachers' report forms used in previous trials was used to examine critically the nature and form of the questions that had been asked. Weightings of various items from questionnaires in the results of the computer analysis helped to discriminate those items which had been most useful from those which had not provided useable information, probably because they were ambiguously worded or difficult to interpret by teachers. Each question was carefully considered in this way, and poor ones were reworded or removed. As a result, the number of questions it was useful to ask was reduced and one form could be produced incorporating all the information which had previously been gathered in three forms for any one unit. In addition, it was possible to combine questionnaires on all four Stage 1 and 2 units of the third set in one form. Admittedly it was a bulky form, but was probably easier to use than many separate ones, and it was certainly more efficient.

THE PREFERENCES FORM

The development of this form has already been described (pages 52 and 53). It was used in the third set trials in exactly the same way as that used for the evaluation of *Early Experiences*. Teachers completed the form, printed on yellow paper, once at a meeting close to the beginning of the trials, and these were returned straight away to the project. At the end of the trials the form, this time printed on green paper, was completed again; the different colours were intended to help the evaluator not to confuse the teachers.

THE DIAGNOSTIC STATEMENTS

In searching for a way of providing some means for teachers to look for progress in their pupils in respect of a wide range of objectives, the list of behaviours—which eventually became called 'diagnostic statements'—was composed. The intention was to help teachers use the observations which they daily make of their children to notice and record whether or not behaviours indicative of the development of certain ideas, skills or abilities were displayed. In the normal course of their work, teachers gather a wealth of information about individual children and can generally give an impressively detailed account of the likes and dislikes, strengths and weaknesses of each one. There is often difficulty, however, in making use of such observations for assessing development of scientific concepts and ideas because teachers do not know what signs to look for. The diagnostic statements provided a list of various things which children may or may not do in relation to science activities, and so tried to help teachers use behaviour indications as pointers to their children's development. The statements were grouped in threes, each group relating to one type of achievement. Where possible there was a progression through the three statements, the first relating to Stage 1, the second to Stage 2, the third to Stage 3, but it was not possible to arrange this throughout the list.

To use the statements teachers were provided with a record sheet, a grid on which children's names were written down the left side and columns provided for checking off statements which observation had shown applied to the children. Teachers were asked to take the booklet page by page, applying each statement to each child before going on to the next statement. A tick was put against the names of children who generally did what was described in the statement being considered, and a blank was left if the statement did not apply in general to a particular child. This was done twice, once at the beginning of the trials and once at the end. By the side of each statement was a list of the reference numbers to one or more of Science 5–13 Objectives for Children Learning Science. Several ways in which these reference numbers might be used were suggested: to indicate objectives to keep in mind for children who had not been ticked for a particular statement; to indicate which objectives may well have been achieved by those children who were ticked, and so suggest progress for them; to indicate a child's stage of development in various aspects of his work and highlight unevenness in development.

FORM FOR COMMENTS ON THE USE OF 'DIAGNOSTIC STATEMENTS'

The 'diagnostic statements' constituted an unfamiliar kind of instrument for teachers, and we knew it was in a rather rough and ready state, so a short one-page form was provided for collecting teachers' reports and comments on its use. It was completed once, after using the diagnostic statements, at the end of the trials.

THE VISITORS' REPORT FORM

Items in the Team Members' Report Form used in the first trials constituted the basis for this form, but there was extensive revision of old items and addition of new ones. The old items were revised in the light of the same kind of scrutiny as used for refining the teachers' report form. The weightings of each item with groups at either end of the main dimensions, in the analyses of earlier trial results, also showed which kinds of information had been most effective in distinguishing positions along the dimension. Items such as 'activities organized so that children can work on their own ploys' and

Title of form	Purpose	When completed	By whom
Report form on Stage 1 & 2 units	To gather opinions, comments, suggestions from teachers about each of the four units	end of trials	teacher
Preferences Form (First)	To indicate teachers' preferences in respect of ways of working in science activities at the start of the trials	beginning of trials	teacher
(Second)	To collect the same information as the above form at the end of the trials and so indicate any changes during the trial period	end of trials	teacher
Diagnostic Statements Record Sheet (First)	To record the teachers' observations of children which would indicate the stage of development of the children at the start of the trials	beginning of trials	teacher
(Second)	To collect the same information as the above form at the end of the trials and so indicate any changes during the trial period	end of trials	teacher
Comments on Diagnostic Statements	To collect feedback on the use of Diagnostic Statements	any time during trials	teacher
Visitors' Report Form	To provide information about the interaction of teacher, pupils and material in the classroom	any time during trials	area representatives or someone acting for him or her

Fig. 14   Programme for completing evaluation forms

'children working at most times individually or in groups at their own tasks' were invariably found to be highly weighted at the end of the dimension indicating satisfactory use of the units. This suggested that more items exploring the interaction between teacher and children would be of value.

Various ways in which classroom interaction could be recorded were once more considered, as before when the team members' form was developed. But again, practical obstacles ruled out methods which required long and detailed records by trained observers. In the circumstances, it was decided to add a new section to the report form, a section which aimed to investigate the work from the children's point of view. To complete this part of the form the visitor was asked to talk with a few children. It was suggested that the teacher should be asked to point out a 'good' and a 'poor' group of children, and asked if she would mind if the visitor talked to each of these groups for a while and looked at what they were doing in some detail. Through doing so, the visitor obtained information about several aspects of the way the children were working, which was recorded immediately after the interview and not in front of the children. The form provided spaces for information from the two groups (indicated by A for 'good', B for 'poor') to be recorded separately.

Figure 14 summarizes the purpose of each of the evaluation forms and indicates the programme for their completion.

Teachers taking part in the trials were also supplied with a short 'Guide to Completing the Evaluation Forms', which explained the purpose of the various forms and gave suggestions for a timetable for completing them.

6 Progress and results of trials of third set and fourth set units

Trials organization for Stage 1 and 2 units of the third set

In the summer of 1970 the number of LEAs officially designated as trial areas was increased by the project in collaboration with the Schools Council. Seven new areas in England were added and four in Scotland. In England the new areas were mainly those in which pre-pilot work had been done—we did not keep to trial areas for this—and their inclusion was an acknowledgement of the help they had already given to the project. In Scotland the increase was part of the dissemination process and only four areas took part in evaluation trials at any one time. The new areas in England each sent a representative to the third meeting of trial area representatives in July 1970, when arrangements for trying the third set of units were first discussed.

The new pattern for trials in which groups of teachers tried groups of units, was proposed at the July meeting and area representatives were given the choice of adopting the new (Trial Pattern 2) or keeping to the previous pattern (Trial Pattern 1). If they chose Trial Pattern 2 they were asked to provide up to four groups, each of four teachers, selected so that regular meetings of teachers within each group would be convenient. For Trial Pattern 1 six classes from each area were requested and area representatives asked to say if they had any preference for trying a particular unit. Decisions about the trial pattern and the choice of classes were made between the third meeting and fourth meeting of area representatives, the latter taking place in December 1970.

Of the nineteen areas in England and Wales, eighteen tried Stage 1 and 2 units, five opted for Pattern 1 and thirteen for Pattern 2. In Scotland four areas were involved, some trying each pattern. Table 3 gives the number of classes beginning and completing the trials according to each pattern. The rate of dropout from the trials was greater than for earlier trials. Two chief reasons can be suggested for this; that teachers had to complete altogether six forms, as described at the end of the last chapter, which was a heavy demand, and that the larger number of classes involved than before in most areas increased administrative problems.

Teachers and administrators from the trial areas attended a one week

Table 3   Numbers of classes taking part in trials of Stage 1 and 2 units of the third set—January to June 1971

| | Trial pattern 1 | | Trial pattern 2 | |
| | Number of classes | | Numbers of classes | |
	starting trials	completing trials	starting trials	completing trials
England & Wales	29	11*	217	165
Scotland	16	16	8	8
TOTALS	45	26	225	173

* The results for one whole area (six classes) were lost in the post and from another area (also six classes) were received after the analysis had been completed.

project course in the holiday immediately before the trials began in January 1971. Through workshops, they became familiar with the new units and the purpose and procedures of the evaluation were explained. The same evaluation forms were used for both trial patterns, which meant that Pattern 1 teachers ignored questions and sections of the teachers' report form referring to the three units which they did not use. The course also provided opportunity for discussion with area representatives about the use of the Visitors' Report Form.

There were no control classes for these trials, so all the teachers involved used one or more units throughout the trial period. This lasted for two terms for Pattern 2 and one term for Pattern 1 trials.

The exact organization of Trial Pattern 2 varied a little from area to area; sometimes there were three or five teachers in a group, but mostly there were four. The location and frequency of meetings also varied locally and teachers would meet either at teachers' centres or in each others' schools, in some cases in each others' homes. The beneficial side effects of these meetings, both for the teachers themselves and for the dissemination of the project's ideas, has already been noted (page 64). In one area the intention of Trial Pattern 2 was misinterpreted and teachers were given the idea that they all had to try all parts of the four units, clearly an impossible task which made the teachers feel under pressure throughout the trials. Despite this and other less extreme variations, the new pattern operated overall as intended and was considered an improvement over Pattern 1 for trials of units of this particular kind.

Results of evaluating the Stage 1 and 2 units

One hundred and ninety-nine classes finished the trial work and completed all the evaluation forms. For each of them data were given in the Teachers'

Report Form, the Visitors' Report Form, the Preferences Forms and the comments on the diagnostic statements. Overall this was a large number of items of information, which was in fact too large to be analysed at one time using the classification computer programme, since the capacity of the computer for this programme would have been exceeded. The analysis of data was therefore carried out in two steps—a course of action which was justified theoretically, though we do not need to go into it here.

INFORMATION NOT SPECIFIC TO ANY UNIT

The first step was the analysis of information which was not related directly to the content of any particular unit but only to the organization of activities, the interaction in the classroom and certain factual data about the teacher and class. This information came from the visitors' report form and the first page of the teachers' report form, where certain background data about the class and teacher were given. The computer programme, which has been described on pages 41 and 42, was used to find groups of items that occurred together. The two groups which were at either end of the chief dimension in the data were easily identified, since they stood out well from the others. The dimension they described was one that had become familiar from earlier analyses: at one end the items indicated a situation in which a teacher uses a discovery method, gives the children responsibility for their work, and puts the project's ideas into practice satisfactorily; items at the other end described a situation in which the teacher directs the children's work, is uncertain or dubious about the project's ideas, and in which the children are highly dependent upon the teacher. The dimension might aptly be called 'grasp and application of the project's ideas'.

The following were the items in the group at one end of the dimension, given in descending order of their weightings in the group together with the percentage of classes to which each item applied (see page 69 for significance of group A and group B children):

Group B children were working on a problem or activity they had suggested themselves 17%

Group B's dependence on the teacher (for ideas, materials, information) was small 9%

Group A children were working on a problem or activity they had suggested themselves 41%

The teacher had warmly approved the project's ideas when first introduced to them 26%

Group A children were recording their work in a way chosen by themselves 26%

Group B children were enjoying what they were doing very much 54%

The class timetable was fully integrated 47%

The children in Group B had a very good grasp of what they were
doing 43%
Group A children were enjoying what they were doing very much 63%
Science activities were carried out at different times as chosen by the
children 43%
The class had previously been used to discovery methods............ 39%
Science activities were organized so children could work on their own
ploys 50%
The children in Group A had a very good grasp of what they were
doing 75%
The teacher used a discovery approach in more areas of the
curriculum 75%
Group B children were making an informal or cooperative record of
their work 73%
The work of Group B was satisfactorily suited to the children's
abilities 81%
Group A children were making an informal or cooperative record of
their work 79%
The work of Group A was satisfactorily suited to the abilities of the
children 89%

These were the items in the group at the other end of the dimension:

Group B children were not much enjoying what they were doing 5%
Group A children were recording their work in a way chosen by the
teacher 20%
Group A children were working on an activity or problem allocated
to them by the teacher 20%
The teacher used a discovery approach hardly at all 7%
Group B children were working on an activity or problem allocated to
them by the teacher 34%
Group A children were not much enjoying what they were doing.... 3%
Group B children were recording their work in a way chosen by the
teacher 23%
The children in Group B had a poor grasp of what they were doing.... 12%
The activities of Group A were too difficult for the children........ 2%
Group B children were highly dependent upon the teacher for ideas,
materials and information 46%
The activities of Group B were too difficult to the children 14%
The headteacher regarded the role of science activities in education as
being useful but not essential 32%
The desks or tables in the room were arranged in regular groups or
rows 31%

These results have been given in full because they provide several kinds of information which was felt to be important. It is recalled that Group A and Group B were the two groups of children interviewed by the visitor to the classroom. Group B were the less able children and it was very interesting that the items concerning them were very highly weighted in the group indicating a good grasp and application of the project's ideas. It seemed to signify that ability to encourage inquiry and independence in learning in less able children was a key factor in discriminating those teachers with a good grasp of the ideas from those with a poor one. Looking at the second group of items gives confirmation for, in this case, the items indicating that the teacher directed the work of even the more able group of children are highly weighted. We investigated this further by looking for other groups provided by the analysis to find items which occurred together with the whole class being below average. These items quite clearly indicated that the teacher directed the work and very little discovery learning was going on. While appreciating that it was very difficult for teachers to avoid this, because slower children tend not to have ideas of their own about what they want to do, the finding was disturbing. We realized that teachers may need special help and ideas for dealing with this problem.

It was also interesting to find that items referring to the interview with the children featured so prominently in the groups. When it was found that results of testing children had not been very effective for the purposes of evaluation, the focus of the evaluation had been turned to the teacher. What was now found was that it was very important to keep the children in focus too, but to look at the way they were working and their immediate grasp of what they were doing rather than at the outcome after a short term trial.

There was considerable value in looking at the degree to which all the items in the Visitors' Report Form were associated with one or other of the groups at the end of the 'grasp and application' dimension. Many of the results of doing this had relevance for rewriting. Others were relevant to decisions which had to be taken by the project team and by others at a later stage—such as how best to introduce the project's ideas to teachers, and what changes in classroom organization and practice might favour the adoption and successful use of the units. Becasue of the interest of the area representatives in these matters a report of the relevant findings was written for them to feed back the information which they may have found useful.

The second step of the analysis of results involved information about each unit. Before this was carried out, however, the results of the first step were used to estimate the position of each class and teacher along the 'grasp and application' dimension. Putting this estimate together with a summary of the information in the visitors' report form enabled the unit authors to begin reading the teachers' report forms, particularly the diary sections, with some idea of the background to the comments and accounts which the teachers had written.

INFORMATION ABOUT EACH UNIT

Part of the data analysed for each unit came the teachers' Preferences Forms. The responses to these forms were scored in the way described in the account of *Early Experiences* evaluation (page 57). The two kinds of information derived from the scores for the computer analysis related to the initial level of score and the change in score; these were indicated by code numbers. At the same time, the overall average pre- and post-trial scores were compared. There was an increase, but it was not large enough to be statistically significant. Further statistical analysis was carried out to see if the overall results concealed changes which were greater for some ranges of pre-trial score than for others, but there were no such changes which could not be explained in terms of trends which can be expected when measurement is repeated. This finding confirmed the results obtained with this instrument from trials of *Early Experiences* that the use of materials per se produced no change in teachers' attitudes to the methods embodied in the materials.

It might be expected that the Preferences scores would be connected with position along the 'grasp and application' dimension. Investigation showed this was the case; correlation of both pre- and post-trial Preferences with a measure of 'grasp and application' were positive and significant. But the coefficients were not very large, which suggested that the two measures had something in common but were by no means measuring the same thing.

Results for each unit came from classes trying both Pattern 1 and Pattern 2 trials. These were not combined in one analysis because the results came from different ways of using the unit and, in any case, the Pattern 1 trials precluded any comparison between the units. As was shown in Table 3, there were twenty-six classes completing trials of Pattern 1. Since this number was divided between four units, the number using any one unit was too small to justify separate analyses for units used in Pattern 1 trials. The procedure adopted for using Pattern 1 results was to summarize the Visitors' Report Form results, giving the position of the teacher along the 'grasp and application' dimension, provide the Preferences results for each teacher, and to pass this information together with the Teachers' Report Forms directly to the unit author. The authors considered this feedback along with results from analysis of Pattern 2 trials.

Among the one hundred and seventy-three classes completing Pattern 2 trials there were some which had to be excluded from the analyses because it was evident that only one unit had actually been used. These results were treated as for Pattern 1. A few others reported on only two or three units but, where it was evident from their diary accounts and comments that they had studied all four units, their results were included in the samples for the units about which they reported. Consequently the samples for the four analyses were not identical.

For the analyses of data about each unit, information from the following sources was coded: the Teachers' Report Forms, the position of the teacher on the 'grasp and application dimension', the initial Preferences score and the change in Preferences score. In the form of code numbers this information was fed to the computer and analysed using the programme for classifying qualitative data. Keeping to what was, by this time, established practice in the project, the evaluator passed the results of the analysis to the team in a confidential report about each unit. Each report gave the items in the groups at the ends of the chief dimension in the data, and the weighting with these groups of other responses about the unit. In all cases, one end of the dimension indicated satisfaction with the unit, and the other dissatisfaction. Comments and criticisms highly associated with dissatisfaction were given most attention because they provided clues to probable reasons for some teachers not finding the unit as helpful as was intended.

Details of the results and interpretations varied for each unit, but there were certain interesting general findings, too. For instance, in all cases the percentage of responses to items in the 'satisfied' categories were very much greater than for the 'dissatisfied' categories. The experience of writing and revising previous units must have accounted for this to a considerable extent. Earlier results had pointed out the kinds of help teachers had found lacking in the first units, and this information had been used directly in writing the third units as well as in revising the first ones. Another general finding was that high initial Preferences scores were strongly associated with satisfaction with the unit. Thus, teachers who used the units successfully tended to be those who already had strong preferences for working through child-centred methods and for providing children with first hand experience.

COMPARISON BETWEEN UNITS

Results of asking for comparisons between the units did not reveal any great preferences for one or another. Teachers were asked in the report form to rank the units according to four different criteria: the amount of use made of the units, their style of suggesting activities, the help they gave in choosing suitable activities, and how much they had stimulated the teacher to continue on her own initiative. The results showed only small differences between rankings given to the units, and all units were given all possible rankings. The information was not helpful for the purposes of rewriting, though it suggested that variety in style of units was perhaps a good thing. Maybe it had been possible to satisfy a wider range of requirements among teachers because of the differences between units than it would have been with four units which were very similar to each other in style.

There was particular interest in finding what light the evaluation results shed upon the value of the different approaches to objectives in the units, which were described on pages 62 and 63. The discussion about objectives in the units was one of the subjects about which opinion was strongly divided. Items relating to the Index of Objectives appeared in the groups at both ends of the dimension for all three units which included an objectives index. It was found that the Index of Objectives was liked and used by those who were generally satisfied with the units and not by those who were dissatisfied. However, the proportions of teachers were very low who had in fact used the Objectives Index very much and found it easy to use. In their comments many teachers pointed out the time it took to look up the reference numbers of the objectives from the list, then look up the pages indicated in the Objectives Index and, when finally having turned to the pages indicated, it was not always easy to identify the relevant activity. Quite evidently, a complex procedure was deterring many from using what might have been a source of help to them.

The approach to objectives in the unit *Coloured Things* was found helpful by a majority of teachers, and no comments on the subject were associated with the 'dissatisfied' group. In *Minibeasts*, the section on 'thinking—through an objective' was thought to be very helpful by sixty-six per cent of teachers, and this response was in the 'satisfied' group of items. Only six per cent thought the section was unnecessary. Other results in the subject showed wide approval of the approach to objectives in this unit.

The use of reference numbers in the margin in parts of the unit *Change* was found confusing by only five per cent of teachers; thirty-one per cent found it very helpful. However, answers to questions about how much use had been made of these numbers revealed that the frequency of use was low. It seemed that these numbers had not increased the reference to objectives as much as had been hoped.

A general finding about objectives was that it was the teachers with a good grasp of the project's ideas who valued help in having the objectives of the activities pointed out, and they did not seem to have a real preference for one approach rather than another. The methods tried were about equally acceptable to teachers who liked to know about objectives, and all more or less equally unacceptable to teachers who thought objectives were irrelevant. As there was no evidence indicating the advantage of any one particular method, there seemed to be no reason why different authors should not maintain their different approaches.

THE DIAGNOSTIC STATEMENTS

The major purposes of providing the diagnostic statements had been to help teachers recognize children's stages of development, and so to match the activities of the children to these stages. As mentioned on page 64, we did not expect the results to be of great value for the revision of the units, and the analysis of results from using the diagnostic statements was put aside because of shortage of time. It was well into the autumn of 1971 before all the results of the third set of trials reached the project, and the end of the project (as far as was known at the time) in the following summer was in sight. Final publication dates were also pressing hard. Consequently the evaluation results had to be analysed, interpreted and used very swiftly, and could not have waited until the 398 record sheets about the children had been processed.

The information from the short form asking for feedback in the use of diagnostic statements showed that sixty-seven per cent of teachers had found the statements of help. From the comments made it was clear that many teachers had been helped, as had been hoped, in choosing the kind of work to suit the children's abilities. On the other hand, while the general idea was approved, the particular statements provided had not been easy to use and were criticized for ambiguity and for involving multiple behaviours. Evidently there was much to be done to produce a form which would be less difficult and time consuming to use. The work involved was much greater than could be tackled at the time but, soon after, the Schools Council set up a project which undertook this work as one of its main concerns. The project, 'Progress in Learning Science', approved in 1971, began at Reading University School of Education, in April 1973, to work on developing material to help teachers observe and record the stages reached by individual children in the attitudes, abilities and concepts relevant to learning science. The project also intends to assist teachers in using information about children's progress in choosing and guiding the children's work.

Using the results

Revision of units in this third set was reported by the authors as being much easier than for previous units. As has already been suggested, experience must have accounted for this to a considerable degree. It is also possible that the information provided from evaluation was more clearly defined and felt to be more reliable than on earlier occasions. The larger number of teachers involved meant reliability was greater on that account alone, but it also meant that a wider range of classes and teachers was sampled. The trials had also run more smoothly than before because of the greater experience of the team and of the area representatives. Finally, the evaluation materials had been refined as a consequence of the earlier trials and the results produced were judged as more valid by the team members. For example, the order in which certain teachers were placed along the 'grasp and application' dimension was said to be exactly what one team member who had visited these teachers in their classrooms had expected. The combination of these various factors increased confidence in the evaluation results.

Evaluation of the Stage 3 units of the third set

The problem of how to help teachers with activities for children at Stage 3, in transition from concrete to formal operational thinking, was the subject of a great deal of discussion and argument both within the team and between the team and various groups of advisers. It was clear that in most classes of children from the age of about ten it was likely that a proportion would have reached Stage 3. The proportion would be small at ten and eleven, increasing at twelve and thirteen, but most optimistic estimates suggested that even at thirteen a substantial proportion would still be at Stage 2 and a few at Stage 1 (see page 47 of *With Objectives in Mind*). Evidently any teacher of Stage 3 children, whether of top age groups in junior schools, or in middle or secondary school, would have many Stage 1 and 2 children in the same class and might want help with activities at all three stages. So it was decided that Stage 3 units should be written on topics that had already been developed at Stages 1 and 2, and that teachers trying out Stage 3 units should be provided with the corresponding Stage 1 and 2 units to use during the trials.

The two Stage 3 units in the third set of units were *Structures and Forces*, Stage 3 and *Change*, Stage 3. The unit *Structures and Forces*, Stages 1 and 2, had been produced in the second set of units and *Change* Stages 1 and 2, was one of the third set. The trials of the Stage 3 units were carried out during the same time as the trials of other units of the third set though they started a little later, in February 1971, and finished in July of that year.

Structures and Forces, Stage 3 had reached printed trial form, but *Change*, Stage 3 was a duplicated edition. In both cases the trials were closer to pilot trials than fully evaluated ones. Feedback was obtained on the unit work by means of a teachers' report form. Information was also gathered on the problems of organization encountered during the trial work by a separate questionnaire focusing upon working conditions. The latter proved helpful in revealing obstacles encountered by teachers in some secondary schools when they attempted to introduce problem solving and active investigation into an organization geared to formal teaching. In addition to these two forms, the teachers taking part in Stage 3 trials were provided with a copy of the diagnostic statements and record sheets for using them. These were supplied in an attempt to help teachers identify the children who were likely to be ready for Stage 3 work and not to provide feedback, so the results were not returned to the project. It was not easy to tell how much help the diagnostic statements were, but it was soon apparent that specialist science teachers did not know their children sufficiently well to use them. These teachers had to use other methods of deciding whether Stage 1, 2 or 3 activities were suitable for individuals. Most began with a variety of activities at all three stages and found the children tended to select their own level if given the choice; one began with all Stage 3 activities on the assumption that his twelve-year-olds must have reached this stage, but soon found that all but a handful were floundering and made better progress with Stage 1 and 2 activities.

Trial areas were given the choice of trying one or other of the Stage 3 units, or neither. Many areas, in the midst of reorganization, preferred not to add to existing confusion in the schools by introducing Stage 3 work at that point, in others teachers of eleven-to-thirteen-year-olds were committed to using a Nuffield Science Programme. Scotland and eleven trial areas in England and Wales did not take part in Stage 3 trials. One area, Southampton, tried only Stage 3 units in its newly established middle schools, and the remaining seven tried Stage 3 units in a varied number of classes and in a variety of kinds of schools. The numbers of classes from the eight areas taking part in the trials were as follows:

> *Structures and Forces*, Stage 3........52
> *Change*, Stage 3 45

Although this may seem a large sample for each unit, it has to be remembered that the unit's activities may only have been used with quite a small proportion of a class. In fact a considerable number of trial teachers found they were not able to make much use of the unit at all because so few children were ready for Stage 3 work. Despite being a rather negative finding this was an important one, emphasizing the importance of selecting children's activities according to their previous experience and achievement, not according to their age.

During the trials the classes were visited by the authors of the two units and by other team members, who fed back their reports directly to the unit authors. At the end of the trials the Teachers' Report Forms were read by the evaluator, but there was such diversity in use of different parts of the units that it was decided there would be no benefit to the authors in summarizing and attempting to interpret the results. The first hand information gained by the authors during their visits enabled them to interpret each teacher's comments in the knowledge of the background to the work. The units were therefore revised by direct application of the feedback provided on the report forms.

In many respects these trials were unsatisfactory, and ideally should have been followed up by further ones, more fully evaluated. The units could not, in short term trials, be used in the way intended. They were written for use with children who had a background of plenty of concrete exploration. This background could not be provided overnight, and trying the units in classes where both teacher and children were new to Science 5–13 materials did not allow their potential value, or their deficiencies, to be thoroughly explored. The problems of setting up trials in classes where Stage 1 and 2 units had been used for a sufficient period beforehand could not be solved in the time available for these trials and called for further work.

Trials of the fourth set of units

At the beginning of the project's fifth year—its final one for producing units—there were four units in various stages of preparation. Arrangements for trials of these units were governed by the necessity to have them completed and the results analysed and used by the end of the year. Short trials were planned for the second half of the spring term and the first half of the summer term and the units were tried in whatever form they had reached by that time.

The amount of previous experience in writing and evaluating units on the part of team members meant that the telescoping of procedures for producing and trying out units, which was forced by circumstances, did not have the detrimental effect upon the end products that it would probably have had in earlier years. In developing the units there was considerable guidance from results of trials of three earlier sets of units and knowledge of the kinds of change which evaluation had shown to be necessary in these units. Continuity in team membership made the impact of this experience the greater. All except one of the authors of these units had been members of the project for at least four years. The one exception was the author of the unit, *Children and Plastics*, but this unit had been in the process of production the longest and already been through informal pilot trials.

In producing the evaluation material there was a similar benefit from the work of previous years. It had been possible to take a critical look at the value accruing from the several different kinds of information which had been gathered in various trials. Shortage of time for the last trials necessitated concentrating resources upon gathering the kinds of information found to be most relevant and easily used. The smoothly running machinery for trials already developed in the areas was also a boon at this time; it was not necessary to brief area representatives on the selection of classes or the completion of the visitors' report forms; they had done it all before.

Not all trial areas were needed for these trials and to have used all of them would have spread our effort too widely, so we concentrated on selected

Table 4 Number of areas and classes involved in the fourth trials

Title of unit	Number of trial areas	Number of trial classes
Ourselves Stages 1 & 2	1	14
Children and Plastics		
Stages 1 & 2	3	40
Like and Unlike		
Stages 1, 2 & 3	3	60
Science, Models and Toys		
Stage 3	3	20

areas. The number of areas and classes involved for each of the units is given in Table 4.

The method and materials used for obtaining feedback on these units varied between them because of the different stages in production they had reached before the trials. It seems better to describe briefly the trials for each unit separately than attempt to generalize about them. It may help to refer back to Fig. 3 on page 22 to recall the various stages in production of a unit.

THE UNIT *Ourselves*

The material in this unit was in the form of suggestions and ideas which were only outlined; it was approximately at the first draft stage in production. The author was using the trials to develop more activities and gather ideas for improving those already suggested. By taking into the unit work which originated in the classroom the viability of the activities in the final unit would be assured.

Information of the kind required was best gathered by the author at first hand; in order to collect it many days were spent in classrooms observing the activities and looking at children's work. The trials of this unit were therefore part of its development, not carried out for the purposes of evaluation. Due to pressure of time the unit went straight from first draft form to final publication.

Science, Models and Toys, STAGE 3

Originally this unit was entitled *Science from Toys*, Stage 3, since it continues the theme of *Science from Toys*, Stages 1 and 2, but the word 'toys' was hardly wide enough to describe the materials used in Stage 3 work. The material in the unit was at second draft stage and the trials were regarded more as pilot trials than as a fully evaluated trials. Visiting by the unit author thus played an important part in interpreting other feedback.

The previous trials of Stage 3 units had provided feedback rather unevenly spread over the units' various sections; some sections were well covered, while others had not been used by many teachers. In these trials the unit author made arrangements for particular sections to be used by different groups of teachers so that in the short time available all the activities in the unit would be tried.

Materials provided for collecting information about the trials were:

a Teachers' Report Form: asking for information about the class and for a report and comments on the unit, section by section

a Visitors' Report Form: the same form as used for trials of the third set of units, completed by the area representative or someone acting for him or her.

As in the case of previous Stage 3 trials it was more useful, especially in view of the short amount of time available, to pass the report forms directly to the author than to try to summarize their contents. The material was revised, using this feedback, and the final edition of the unit produced.

Like and Unlike, STAGES 1, 2 AND 3

Activities for all children at all three stages were suggested in this unit, the only one to attempt the combination. The age range in the classes in which it was tried therefore had to be greater, and not all parts of the unit were used by any one teacher. The material was at second draft stage and first hand observation of the work in progress by the unit author was an essential part of gathering information for revising the unit. Other feedback came from two forms: a Teachers' Report Form on the unit, and the Visitors' Report Form completed by the area representative. The information on these forms was not summarized, but used directly by the author for revising the material and producing the final version of the unit.

Children and Plastics, STAGES 1 AND 2 AND
BACKGROUND INFORMATION

This was the only unit of the fourth set to have gone through all the phases of development which were usual for units in earlier sets (see page 22). The activities in the unit had been given limited pilot trials, but we wanted to find out how helpful the unit as a whole, including the background information, was to a wide range of primary teachers. The evaluation data had to be gathered as efficiently as possible so that it could be analysed and used within a matter of a few weeks after the end of the trials.

Four kinds of information were gathered:

(a) information about the class, teacher and school,
(b) the teachers' report and opinion about the unit and the work arising from its use,
(c) information about the interaction of teachers, children and materials, and
(d) information about the school environment and learning environment within the school.

These were collected jointly on two forms: a report form completed by the trial class teacher, and a report form completed by someone who visited the class (either a team member or the area representative or someone acting for him or her). In addition each class was visited by the author of the unit, if possible, at least once during the trials, so that she gained impressions of the work in action at first hand.

Work was completed in all forty classes which began the trials, but one Teachers' Report Form was not received in time for the analysis. Processing results from the Visitors' Report Form was rapid because it was possible to make use of the results of analysing a large amount of data from the same forms in the third trials. The previous analysis indicated the weighting of all the items in the form with the groups at the end of the dimension which had been called 'grasp and application of the project's ideas'. By applying these weightings, the information on the forms for the fourth trials was used to place each teacher and class along the same dimension. The teachers were placed in one of three broad bands indicating a very good, moderate or not very good 'grasp and application'. A code number signifying this information was added to the coded data from the Teachers' Report Form and fed into the computer. The programme for classifying qualitative data, familiar from previous trials, was used to analyse the data for each class.

The two groups representing the ends of the chief dimension in the data are given below in descending order of weighting in the group, with the number to which each applied (out of a maximum of thirty-nine). The items in the group at one end were:

The section 'Some Common Plastics' should be made longer by including more examples .... (8)
The Background Information was used by reading sections as required during the work ....(13)
The reference books suggested were found useful and sufficient......(11)
Suggestions as to how the work could lead to activities in other areas of the curriculum were thought to be adequate....................(26)
The Background Information was thought to be very useful ....(20)
The teacher was satisfied with the active response of the children to the work ....(21)
None of the activities suggested was found unsuitable in practice....(26)
The unit gave adequate help with starting the work going ....(34)

Those in the group at the other end were:

The number and variety of activities suggested in the unit was too great to be helpful .... (3)
The guidance given as to what activities are appropriate to Stage 1 and Stage 2 was confusing .... (1)
The section 'What Are Plastics?' was very difficult to understand.... (1)
As a result of the work hardly any children were able to identify common plastics .... (6)
The teacher was not satisfied with the active responses of the children to the work .... (5)
There was not enough direct guidance for teachers given by the unit.. (9)
Suggestions for how the work could lead to activities in other areas of the curriculum were thought unnecessary .... (8)

The classes in the school were partially streamed .... (3)
Some of the activities suggested were found too difficult for any of the
children .... (9)
Some of the activities carried out were found unsatisfactory in practice. .(13)
In general the Background Information was found quite (but not very)
useful ....(14)
No changes considered necessary in the section 'Some Common
Plastics' ....(26)

The degree of association with each of these groups was found for all
items in the Teachers' Report Form from the weightings provided by the
computer print-out. As found in previous similar analyses it was clear that
items in, or strongly associated with, one group indicated broad satisfaction
with the unit, and in the case of the other group indicated dissatisfaction.
Also, a not very good 'grasp and application' was strongly associated with the
second group above, and a very good 'grasp and application' with the first
group.

Frequencies of response in 'satisfied' categories were much greater than in
'dissatisfied' ones, showing that the unit was generally acceptable to a majority
of the trial teachers. However not all teachers had been able to use the unit
in a way that judged to be successful, and the reasons for this were sought
in the responses associated with dissatisfaction. It was found that three main
kinds of help were requested: with the management of activities, with selection
of activities for younger and less able children and with the details of certain
activities. Teachers' free comments enlarged upon these points and indicated,
for instance, which activities required further details. Criticisms and suggestions about other aspects of the unit, the Background Information, safety
precautions, reference books, sources of materials, and so on, were also
obtained in detail.

The findings from the analysis were conveyed to the unit author in a report,
and the author also read the teachers' accounts of their work. The report
on the unit, slightly modified, was later reproduced and became the fourth
evaluation report for teachers.

7 Evaluation of the project: past, present and future

It is fitting that a project which endorses and promotes discovery learning should have provided opportunity for learning, through discovery, a great deal about the effectiveness of various approaches to the evaluation of its materials. Because Science 5–13 produced and tried out its units in separate sets it was possible to learn from experience the importance of gathering information of various kinds to suit the aims of the formative evaluation. Much was 'learned by doing' about the types of information which can be effectively used for particular purposes, and about ways of gathering them.

While it is not the intention to summarize here what has been said about the evaluation in earlier chapters, it seems relevant to point out some of the lessons which have been learned. Undoubtedly the most positive aid to revising the trial units came from the teachers' questionnaires; here were the reports of how far the suggestions could be put into practice, the accounts of how activities had begun and ended, reasons for parts of a unit being unsatisfactory, and suggestions for changes. No other source of information was so rich in detailed and definite indications of how the material could be made more helpful to teachers. Yet it must be said straight away that the value of this information would have been far less without the addition of data from observation of how the units were being used. Furthermore, the use of a computer programme to classify information from teachers and classroom observers added significantly to the usefulness of such data. Without this facility there would have been ambiguities in interpreting teachers' comments, arising from inability to relate them systematically to the background in which they had been made.

From the first set of trials it was learned that information coming from children's test results was tentative and not readily usable for guiding rewriting without being supplemented by other data. The results played a useful part in confirming that the general approach of the material was effective in promoting achievement of its stated objectives, and the development of tests also had side benefits for the production of units. But for indicating changes which would make the units more effective they were of much less use than information from other sources. The tests were also by far the most expensive item in the evaluation, both in direct cost and in man/woman hours. While it could not be said that the test information was without value for this project, it can be said that, where resources are limited and it is necessary to concentrate upon gathering information to give the greatest return on money, time and human energy, then the choice would be for teachers' reports and direct observations in the classroom and not for tests of short term changes in children's behaviour.

Results for the second set of units largely confirmed this conclusion about the value of test results in relation to other data. In the evaluation of *Early Experiences* the additional information provided by using the Preferences Form indicated an important relationship between teachers' attitudes and satisfactory use of the unit. Teachers with a high score on attitude towards giving children freedom to learn through active exploration tended to be those who used the activities as intended, while the reverse was true for those with a low score. No significant change in attitude was detected during the trial period, a finding later confirmed from the more extensive use of the Preferences Form in the third trials.

The findings from earlier trials made it possible to learn about the content of the evaluation instruments as well as about sources of information. The questions asked in the teachers' forms used in the first two sets of trials were examined critically to discover if the answers had been used at all in making decisions about revising the unit. Some questions were eliminated as a result; for example, feedback from an item about the desirability of having a subject index and from another about the value of photographs in the unit had not been used because decisions about these things were based on other considerations. Questions of which answers had been used were examined for overlap and ambiguity; the results of the computer analysis were invaluable for this. After the reduction and modification of questions in the forms, which then became possible, new items were added to draw out information which the authors had felt lacking.

The results of the classification analysis were similarly used to reshape the team members' report form, used in the first and second trials, into the visitors' report form, used in the third and fourth trials. The purpose of these forms was to supply information about how the units were being used, so it was necessary for the forms to be efficient, to focus the observations on features and events which were particularly pertinent to the use of the units, and not to consume the observers' time in recording things common to most classrooms. Among the items most highly weighted in the groups from the analyses of earlier results were the few relating to contacts between teachers and children. It was felt there were not enough of these, so a new section was added to the Visitors' Form in an attempt to investigate more closely the interaction between teachers and children, and do this from the children's point of view. Findings from extensive use of the resulting form led to another discovery; information coming directly from talking to the children was extremely effective in describing how the material was used. Furthermore, looking at the way in which less able children were working was especially helpful in making discriminations according to the 'grasp and application of the project's ideas'.

Throughout the various trials the results underlined the value of gathering and combining information about different aspects of the use and effect of the units. Information from any one source was inadequate on its own, and really became useful only when supported by evidence from other sources. In describing the various formal evaluation instruments it should not be forgotten that information was also gathered informally by the team members in their contacts with teachers and visits to trial classes. In general the formal evaluation results gave substantial backing to the informal impressions gathered by the team members. This was important, for it would have been disquieting if one had led to a very different picture of the trial situation than the other. The formal results were used with greater confidence because they supported, and were supported by, what the team felt. Nevertheless the team's impressions were gathered from short and probably unrepresentative glimpses of the work, and the evaluation provided a more thorough account of what went on throughout the trials.

So much for the past, but what has been learned from the evaluation which is of present value? First, it is hardly necessary to state that the units in their final form are likely to be more helpful to more teachers because of the information provided by the evaluation. It is also possible that teachers may use the units with greater confidence, knowing that they were widely tried out and evaluated during their production.

A second point refers to application of what was found about the combination of circumstances associated with using the unit as intended. Repeatedly, in different trials, the same features were found to be present when the units were used with success and brought satisfaction to the teachers. Some of these features, such as those relating to the school environment and the past experience of the teacher and children, could not now be affected, but others —at least in theory—were susceptible to change. The latter provided detailed indications of the changes in classroom practice that might profitably be made. The adoption of conditions favourable to optimal use of the units was urged in the revised version of *With Objectives in Mind* and in the units, but it was also kept in mind as a major aim of courses preparing teachers for using the units.

Details of these favourable conditions were communicated to the area representatives, as mentioned on page 74, because these were the people responsible for running courses in various parts of the country, not only for the trials but also for the wider dissemination of the project's materials. It seems a reasonable conjecture, moreover, that taking part in gathering information on the Visitors' Form may have helped the area representatives to identify critical aspects of classroom practice relevant to successful use of the materials. Consequently they may have been able to direct help in courses to these aspects.

The project is making a more direct attempt to help teachers and course leaders when the project's materials are introduced for the first time. The unit *Understanding Science 5–13* is designed to help the understanding of the project's philosophy, its aims and objectives, the teacher's role in using the units, the interaction of objectives and activities and the application of all these ideas in the classroom. It consists of a guide in the form of a programmed book, supported by slides and tapes, and can be used in a number of ways, as material for a course or for self-instruction.

The impetus to produce this unit came largely from contrasting the way units were used by teachers who had been on a course, and those who used the units without the benefit of a course. The unit is an acknowledgement that the project's books do not on their own give as much help as many teachers need in starting children learning science actively for the first time. The evaluation findings confirmed, with objective measurement of teachers' attitudes (see pages 58 and 75), that using the units did not significantly change teachers' willingness to adopt methods for enabling children to learn more active inquiry. Although increase in attitude score was associated with successful use of the units, these increases were among those who already had a favourable attitude. Here was further evidence that some action was needed to improve attitudes before the materials were used.

No further evaluation of Science 5–13 is planned at present. Nevertheless, it may be useful to review the outstanding problems and suggest what seem to be, at this time, the more important subjects for later evaluation and research.

At many points in this report shortage of time has been mentioned as a restraining factor; it has often determined limitations on how problems could be approached and even which problems could be tackled. In particular it prevented any long term study of the use of the material and any longitudinal survey of the effect of the material on the children. This is seen as a most important subject for future work. It may have value not only with respect to Science 5–13, however, since it could help to validate the evaluation strategy. In the third and fourth trials evaluation attention shifted from the outcomes of using the materials to the processes by which the outcomes were expected to be brought about. The changed strategy proved more effective for formative evaluation but involved the assumption that, if the children's learning is conducted in a certain way, then intended outcomes will follow. There needs to be some test of the truth of this assumption. Such an investigation might try to correlate long-term changes (over several years) in children's behaviour with various different approaches, one of which would be that of Science 5–13, to producing desired changes.

For the above investigation to be possible it is assumed that teachers will use Science 5–13 materials for some years to come and that at least some of them will use the material as intended. However, this assumption may not be true, and introduces another set of questions which should be systematically studied. Hopefully, by the end of the next three to five years there will be some teachers using Science 5–13 materials, some who took them up but dropped them, some who never heard about them, and some who heard about them but were not interested. To inquire about reasons for various patterns of usage or non-usage would be valuable in many ways. It could throw further light upon the problems of dissemination, on the success of various in-service courses, and provide extremely useful data for the next wave of curriculum development in science for five-to-thirteen year olds. It may be possible, for instance, to distinguish varying degrees in the application of the project's philosophy—from using the units as vehicles for putting the philosophy into practice to using the units as ideas for teaching in a way quite out of keeping with the philosophy. If failure to apply the philosophy is widespread, it may mean that it is unrealistic in the changing school situation, and new approaches may have to be developed. On the other hand, teachers may become more able to apply the philosophy as they become more confident in guiding the activities. We do not know which of these, or variations of them, will be the case; we should find out.

Further work is also required in relation to development of units. It may be useful to produce material for pre-school children, especially now that there will be more of them. Then there is the relation between mathematics and science activities for young children. Does it make sense to continue developing science and mathematics curriculum material separately?

At the top of the age range the remaining problems abound. It was noted on page 80 that the trials and evaluation of Stage 3 units had been in many respects unsatisfactory. There is especial need for long term study of the problems of providing suitable learning experiences in science for children between eleven and thirteen. What Science 5–13 has been able to do so far has only been a first step, made somewhat in the dark because of the distinct and widespread absence of detailed information about the way children in this age group learn. There is room for much research on this subject, but following the progress for several years of children whose teachers use Science 5–13 units might also make a significant contribution to knowledge about it.

It would not be useful to suggest that any of these problems have priority, since their relative importance will depend very much upon what happens to Science 5–13 materials in the next few years. This may in turn be strongly affected by whether some organization is maintained for servicing the project's materials when the team finally breaks up. However, it is not inappropriate to recommend that some future work should be done, if only because—to echo a remark at the beginning of this report—Science 5–13 was a major Schools

Council project, and therefore an expensive undertaking. It will provide opportunity for exploring many of the long term problems involved in inducing curriculum changes which are relevant to other projects. It would make sense, educationally and financially, to take advantage of this opportunity.

Appendix A
Units and evaluation materials

Units in the four trial sets

	With Objectives in Mind	Guide to Science 5–13
First set:	*Metals*	Stages 1 and 2
	Metals	Background Information
	Working with Wood	Stages 1 and 2
	Working with Wood	Background Information
	Time	Stages 1 and 2 & Background
	Trees	Stages 1 and 2
Second set:	*Early Experiences*	Beginnings
	Structures and Forces	Stages 1 and 2
	Science from Toys	Stages 1 and 2 & Background
Third set:	*Change*	Stages 1 and 2 & Background
	Minibeasts	Stages 1 and 2
	Coloured Things	Stages 1 and 2
	Holes, Gaps and Cavities	Stages 1 and 2
	Structures and Forces	Stage 3
	Change	Stage 3
Fourth set:	*Children and Plastics*	Stages 1 and 2 & Background
	Ourselves	Stages 1 and 2
	Like and Unlike	Stages 1, 2 and 3
	Science, Models and Toys	Stage 3

A resource unit *Using the Environment*, by Dr Margaret Collis, has also been produced in four parts:

Early Explorations
Investigations (2 volumes—Part 1 & Part 2)
Tackling Problems
Ways and Means (2 volumes—Part 1 & Part 2)

Understanding Science 5–13 is a unit which introduces the project's materials for those to whom the ideas are new.
Publisher for the project—Macdonald Educational

Evaluation materials used in the trials

TRIALS OF FIRST SET OF UNITS

For all units:	Visit Report Form for team members
	Teachers' Report Form A
	Teachers' Report Form B
	Guide to teachers administering the group tests
	Six General Test film loops
	General Test film commentary
	General Test children's answer booklets
	General Test marking scheme
	General Test mark record sheets
Working with Wood	Teachers' Report Form C
	Working with Wood test film loops (four)
	Working with Wood test commentary
	Working with Wood test children's answer booklets
	Working with Wood test marking scheme
	Mark record sheets
Metals	Teachers' Report Form D
	Metals test film loops (four)
	Metals test commentary
	Metals test children's answer booklets
	Metals test marking scheme
	Mark record sheets
Time	Teachers' Report Form E
	Time test film loops (four)
	Time test commentary
	Time test children's answer booklets
	Time test marking scheme
	Mark record sheets
Trees	Teachers' Report Form F
	Trees test film loops (five)
	Trees test commentary
	Trees test children's answer booklets
	Trees test marking scheme
	Mark record sheets

TRIALS OF SECOND SET OF UNITS

For all units:	Visit Report Form for team members
	Teachers' Report Form A
	Teachers' Report Form B
Structures and Forces	Teachers' Report Form H
Science from Toys	Teachers' Report Form G

Early Experiences	Teachers' Report Form I
	Teachers' Preferences Form
	'Selected activities' individual tests for children
	Teachers' observation of children's attitudes

TRIALS OF THIRD SET OF UNITS

For all units:	Visitors' Report Form
For all Stage 1 and 2 units:	Teachers' Report Form on Stage 1 and 2 units
	Diagnostic statements
	Record sheets for diagnostic statements
	Report sheet on diagnostic statements
	Teachers' Preferences Form
	Notes for teachers on the evaluation materials
For all Stage 3 units:	Report form on working conditions
	Diagnostic statements
	Record sheets for diagnostic statements
Structures and Forces Stage 3:	Teachers' Report Form on *Structures and Forces* Stage 3
Change Stage 3:	Teachers' Report Form on *Change* Stage 3

TRIALS OF FOURTH SET OF UNITS

For all units:	Visitors' Report Form
Children and Plastics:	Teachers' Report Form on *Children and Plastics*
Science, Models and Toys Stage 3:	Teachers' Report Form on *Science, Models and Toys* Stage 3
Like and Unlike:	Teachers' Report Form on *Like and Unlike* Stages 1, 2 and 3

Appendix B
Committees of the project

Consultative committee

L. C. Comber—Chairman, formerly Staff Inspector for Rural Science, Department of Education and Science

F. F. Blackwell, Inspector of Schools, Croydon Education Committee

N. Booth, HMI, Staff Inspector for Science, DES

P. Boyers, Senior Lecturer, Coventry College of Education

M. A. Brimer, Head of Research Unit, Bristol University School of Education

Mrs G. Dawes (from September 1972), Headmistress, Two Mile Hill Infants' School, Bristol

Dr J. Duffey, formerly Chief Inspector of Schools, Bristol Education Committee

M. J. Elwell, formerly Organizer, Nuffield Combined Science Project

M. Harris, formerly Director, Environmental Studies Project

Miss M. J. Hay, HMI, Scottish Education Department

Dr P. J. Kelly, formerly Joint Organizer, Nuffield A-level Biology Project

Professor K. Keohane, Co-ordinator, Nuffield Science Teaching Project Director, Centre for Science Education, Chelsea College of Science and Technology

E. J. Machin, Chief Inspector, Birmingham Education Committee

D. H. J. Marchant, Chairman, Primary Science Committee, Association for Science Education

Professor G. Matthews, Organizer, Nuffield Mathematics Teaching Project

Mrs H. Misselbrook, formerly Organizer, Nuffield Secondary Science Project

A. J. Rose, Headmaster, Shaftesbury Junior School, Leicester

Professor A. Ross, Department of Educational Research, University of Lancaster

Professor W. Taylor, Director, Bristol University Institute of Education (at that time)

Miss R. M. Tirkin (deceased August 1971), Headmistress, Four Acres Infants' School, Bristol

E. R. Wastnedge, HMI, formerly Organizer, Nuffield Junior Science Project

Miss N. Bartman (from February 1973), Project Officer, Schools Council

Dr T. Burdett, HMI (from September 1972), Educational Adviser, Science, Schools Council

P. S. Clift (September–December 1971), Project Officer, Schools Council
Dr H. Jones, HMI (September 1967–February 1969), Educational Adviser, Science, Schools Council
G. Porter (January 1972–February 1973), Project Officer, Schools Council
R. D. Price (September 1967–August 1971), Project Officer, Schools Council
Dr C. Selby, HMI (February 1970–August 1972), Educational Adviser, Science, Schools Council

Trial area representatives

Anglesey	W. J. Williams	County Primary School, Valley, Holyhead, Anglesey
Birmingham	E. J. Machin	Education Department, The Council House, Margaret Street, Birmingham B3 3BU
	F. Darrall Head of Department Primary Maths and Science	Science and Maths Centre, St. Luke's Road, Birmingham B5 7DA
Bradford	R. I. Shewell, succeeded by B. Shillaker Science Adviser	Education Department, City Hall, Bradford 1
Bristol	A. J. Light, succeeded by A. McFarland Inspectors of Schools	Education Department, The Council House, College Green, Bristol BS1 5TN
	G. Day Warden	Hannah More Teachers' Centre, New Kingsley Road, St. Phillips, Bristol BS2 0LT
Cardiff	Miss B. G. Sneyd, succeeded by B. Grady Science Adviser	Education Department, Municipal Offices, Kingsway, Cardiff CF1 4JG
Carlisle	G. McAdam Headmaster	Belah Primary School, Eden Street, Carlisle CA3 9JZ
Croydon	W. A. C. Bullock Inspector of Schools	Education Department, Taberner House, Park Lane, Croydon CR2 1TP
Essex	D. T. Griffiths Science Adviser	Education Department, County Hall, Chelmsford, Essex

Gloucestershire	A. Garnham Assistant Education Officer for Curriculum Development & In-service Education	Education Department, Shire Hall, Gloucester GL1 2TP
ILEA	Miss D. Alexander, succeeded by G. Edwards Teacher/Adviser	Kingswood Science Centre, Kingswood Primary Boys' School, Gipsey Road, London S.E. 27
	T. Carter Warden	North London Science Centre, 62–66 Highbury Grove, London N. 5
Kent	Dr M. Collis Inspector of Schools	Education Committee, Springfield, Maidstone, Kent
Leicester	R. J. Pickering succeeded by G. Over Warden	Science Centre, Crown Hills House, Gwendolen Road, Leicester LE5 5FP
	M. J. Perry Science Adviser	Education Committee, Newarke Street, Leicester LE1 9BF
Liverpool	G. D. Williams	Mosspits Lane C.P. School Junior Mixed Department, Liverpool L15 6UN
Somerset	Dr J. Taylor, succeeded by Dr P. Armitage Science Adviser	Education Department, County Hall, Taunton, Somerset
Southampton	N. M. Griffiths, succeeded by D. Evans	Education Department, Civic Centre, Southampton SO9 4XE
Stafford	K. Wild Science Advisory Officer	Education Committee, County Education Office, Stafford
St. Helens	J. E. Watson Science Adviser	Education Department Century House, St. Helens, Lancs
Teesside	J. Milbourn Science Adviser	Education Offices, Woodlands Road, Middlesborough, Teesside TS1 3BN

| *West Riding* | Miss J. P. Imrie Area Adviser | County Education Office, Bond Street, Wakefield, Yorks |
| *Scotland* | Miss M. J. Hay, HMI | 22 St. Clair Terrace, Morningside, Edinburgh EH10 5NW |

Appendix C
Reports, papers and articles relating to Science 5–13 or to its evaluation

L. F. Ennever, 'Helping Children to Learn Science', *Dialogue* (Schools Council Newsletter) No. 1, September 1968, 5–6

L. F. Ennever, 'Science 5–13', *Education and Science* No. 33, June 1969

Science 5–13 *Newsletter 1*, June 1969*

L. F. Ennever, 'The New Science', in *Teaching in the British Primary School*, ed. Vincent R. Rogers. Macmillan, 1970

'Early Experiences, Science 5–13', *Dialogue* (Schools Council Newsletter) No. 6, August 1970, 12

W. Harlen, 'Some practical points in favour of curriculum evaluation', *Journal of Curriculum Studies*, **3**, 2, November 1971

W. Harlen, *Report for Teachers on the Evaluation of the First Set of Units,*1971*

W. Harlen, *Report for Teachers on the Evaluation of the Second Set of Units,* 1971*

Science 5–13 *Newsletter 2*, February 1971*

L. F. Ennever, 'Science 5–13', *Dialogue* (Schools Council Newsletter) No. 11, May 1972, 3–4

W. Harlen, *Report for Teachers on the Evaluation of the Third Set of Units,* 1972

W. Harlen, *Report for Teachers on the Evaluation of 'Children and Plastics', Stages 1 and 2, & Background Information,* 1972

W. Harlen, 'Formulating objectives—problems and approaches', *British Journal of Educational Technology*, Vol. 3, No. 3, October 1972

W. Harlen, 'Science 5–13', in *Evaluation in Curriculum Development: Twelve Case Studies.* Schools Council Research Studies, Macmillan Education 1973

W. Harlen, 'The effectiveness of procedures and instruments for use in formative curriculum evaluation', (unpublished PhD thesis). Bristol University, 1973 (2 vols).

* Out of print.

Appendix D
Information about participants
in the unit trials

Information about classes taking part in trials of the first set of units

Areas	Number of classes				Unit tried	Duration of trials
	Trial		Control			
England & Wales						
Anglesey	4	(4)	4	(4)	*Working with Wood*	October 69–March 70
Birmingham	7	(7)	7	(7)	*Working with Wood*	October 69–March 70
Bristol	5	(6)	5	(6)	*Time*	October 69–March 70
Cardiff	6	(6)	6	(6)	*Trees*	October 69–July 70
Carlisle	6	(6)	6	(6)	*Metals*	October 69–March 70
Essex	6	(6)	6	(6)	*Time*	October 69–March 70
Kent	5	(6)	5	(6)	*Trees*	October 69–July 70
Leicester	5	(6)	5	(6)	*Trees*	October 69–July 70
Liverpool	6	(6)	6	(6)	*Metals*	October 69–March 70
London (ILEA)	6	(6)	6	(6)	*Time*	October 69–March 70
St. Helens	6	(6)	6	(6)	*Metals*	October 69–March 70
West Riding	6	(6)	6	(6)	*Working with Wood*	October 69–March 70
Scotland						
Dundee	2	(2)	2	(2)	*Time*	October 69–March 70
Lanarkshire	2	(2)	2	(2)	*Time*	October 69–March 70
Roxburghshire	2	(2)	2	(2)	*Time*	October 69–March 70
West Lothian	2	(2)	2	(2)	*Time*	October 69–March 70
TOTALS	76	(79)	76	(79)		

Note: Bracketed numbers began, but did not complete, trials.

General information about the trial classes for the first set of units

	Metals	Working with Wood	Time Eng*	Scott†	Trees	TOTALS
Type of school						
Junior with Infants	7	14	9	7	5	42
Junior Mixed only	10	3	7	0	11	31
Junior Boys only	1	0	0	0	0	1
Junior Girls only	0	0	0	0	0	0
Size of school						
8 or more classes	13	9	10	7	13	52
Fewer than 8 classes	5	7	6	0	3	21
School catchment area largely						
Urban	7	8	5	2	7	29
Suburban	11	2	13	4	9	39
Rural	1	8	2	2	2	15
(two of these were ticked if there was a mixture)						
Background of the children generally						
Prosperous	5	1	4	2	2	14
Average	9	12	11	6	9	47
Disadvantaged	5	5	3	0	6	19
(two of these were ticked if there was a mixture						
Approximate age of School building						
Less than 15 years	9	2	1	5	6	23
15 to 50 years	4	4	10	0	3	21
More than 50 years	5	11	5	2	7	30
The school grounds contained						
Large areas of grass	14	8	12	3	9	46
Trees	9	9	13	2	9	42
Flower beds and borders	9	7	11	2	9	38
Garden plots for children's use	0	3	2	0	2	7

* Totals for *Time* in England and Wales are reduced by one because one teacher did not complete the questionnaire although all the children's tests were completed.

† Totals for *Time* in Scotland are all reduced by one because one teacher did not complete the questionnaire although all the children's tests were completed.

List of schools involved in trials of the first set of units

England and Wales

ANGLESEY

Trial
Llanfawr County Primary School
Newborough County Primary School
Pencarnisiog County Primary School
Llanddeusant County Primary School

Control
The Thomas Ellis Voluntary Primary School
Menai Bridge County Primary School
Aberffraw County Primary School
Llanedwen Voluntary Primary School

BIRMINGHAM

Trial
Billesley Junior School
Chandos Junior & Infant School
Colmore Junior School
Pineapple Junior and Infant School

Control
Tile Cross Junior and Infant School
Foundry Junior School
Perry Common Junior School
Park Hill Junior and Infant School

BRISTOL

Trial
Whitehouse Junior Mixed School
Four Acres Junior Mixed School
Henleaze Junior Mixed School
Westbury Park Junior Mixed School
Bridge Farm Primary School
Ashley Down Junior Mixed School

Control
Embleton Junior Mixed School
Fonthill Junior Mixed School
Elmlea Primary School
Summerhill Junior Mixed School
Stoke Bishop C.E. Primary School
Tyning Junior Mixed School

CARDIFF

Trial
Peter Lea Junior School
Kitchener Junior School
Gladstone Junior School
Penybryn Junior School
Glanyrafon Junior School

CARLISLE

Trial
Norman Street Primary School
Kingsmoor County Primary School
Stanwix Primary School
Upperby Primary School
Pennine Way Junior School

Control
Belah Primary School
Belle Vue Primary School
Greystone Junior School
Inglewood Junior School
Caldewgate Junior School
Denton Holme Junior School

ESSEX

Trial
Deneholm County Primary School
Ockenden Shaw County Junior School
Orsett C.E. Primary School

Control
The Tyrrells County Primary School
Trinity Road County Primary School
The Tyrells County Primary School
Great Easton C.E. School

ILEA

Trial
Crown Lane Primary School
Fenstanton Junior School
Sunnyhill Junior School

Rosendale Junior School
St. Andrews Stockwell C.E. School

Control
Granton Primary School
Jessop Primary School
Eardley Junior School
Norwood Park Primary School
Julians Primary School

KENT
Trial
Selstead C.E. Primary School
Christchurch C.E. Primary School
Seabrook C.E. Primary School
Cheriton County Primary School
Stella Maris R.C. Primary School
Mundella County Primary School

Control
St. Martins C.E. Primary School
Harcourt County Junior School
Sandgate C.E. Primary School
All Souls C.E. Primary School
Willesborough County Junior School
St. Peter's C.E. Primary School

LEICESTER
Trial
Hazel Junior School
Overdale Junior School
Braunstone Frith Junior School
Shaftesbury Junior School
Forest Lodge Junior School
Whitehall Junior School

Control
Hazel Junior School
Overdale Junior School
Braunstone Frith Junior School
Shaftesbury Junior School
Forest Lodge Junior School
Whitehall Junior School

LIVERPOOL

Trial

Woolton County Primary School
Joseph Williams County Primary School
Broadgreen County Primary School Boy's Department
Kingsthorne County Primary School
Craighurst County Primary School

Control

Northway County Primary School
Dovedale County Primary School
Stockton Wood Road County Primary School
Mosspits Lane County Primary School

ST. HELENS

Trial

Allanson Street County Primary Junior School
St. Anne's Junior Mixed School
Carr Mill Junior School

Control

Thatto Heath Junior Mixed School
St. Matthews C.E. School
Parr Flat Junior Mixed School
Windlehurst County Primary

WEST RIDING

Trial

Cowling County Primary School
Earby Kelbrook County Primary School
Grassington C.E. School
Bradleys Both County Primary School
Skipton Parish C.E. Primary School
Skipton Ings County Primary School

Control

Kiddwick C.E. School
Salterforth County Primary School
Carleton (Skipton) Endowed School
Cononley County Primary School
Silsden Aire View School
Skipton Greatwood County Primary School

S. 5-13.—5*

Scotland

DUNDEE

Trial
St. Clements Primary School
Eastern Primary School

Control
St. Ninian's Primary School
Forthill Primary School

ROXBURGHSHIRE

Trial
Kelso Primary School
Channelkirk Primary School

Control
Kelso Primary School
Merton Primary School

LANARKSHIRE

Trial
Tollbrae Primary School
Townhill Primary School

Control
St. Edwards R.C. Primary School
Ladywell Primary School

WEST LOTHIAN

Trial
Windy Knowe Primary School
Boghall Primary School

Control
Kirkliston Primary School
Balbardie Primary School

Information about classes taking part in the trials of the second set of units

Areas	Science from Toys		Structures & Forces		Early Experiences Trial		Control	
England & Wales								
Anglesey			1	(3)	2	(3)		
Birmingham			6	(6)	5	(5)		
Bristol			5	(5)	4	(4)	4	(4)
Cardiff			6	(6)	4	(4)		
Carlisle	6	(6)			4	(4)		
Essex			6	(6)	4	(4)		
Kent	4	(6)			4	(4)	4	(4)
Leicester	6	(6)			4	(4)		
Liverpool	6	(6)			4	(4)		
ILEA			3	(4)	3	(3)		
St. Helens	1	(1)			1	(1)		
West Riding			3	(6)	3	(4)		
Scotland					2★	(2)		
TOTALS	23	(25)	30	(36)	45	(47)	8	(8)

★ See page 112.
Note: Bracketed numbers began, but did not complete, trials

General information about the trial classes for the second set of units

	Structures and Forces	Science from Toys	Early Experiences		TOTALS
			England	Scotland	
Type of shool					
Junior Mixed only	15	8			23
Junior with infants	15	15	20	2	52
Infants only			23		23
Size of school					
8 or more classes	21	18	18	2	59
Fewer than 8 classes	9	5	25		39
School catchment area largely*					
Urban	8	8	8		24
Suburban	18	12	30	2	62
Rural	11	4	8		23
Background of the children generally*					
Prosperous	7	5	5	1	18
Average	23	15	29	1	68
Disadvantaged	7	6	9		22
Sex of teacher					
Male	23	9	1		33
Female	7	14	42	2	65

* Two of these were ticked if there was a mixture.

Lists of schools completing trials of the second set of units

England and Wales

ANGLESEY—ALL TRIAL CLASSES

Rhosneiger County Primary School
Llanfawr County Primary School
Newborough County Primary School

BIRMINGHAM—ALL TRIAL CLASSES

Lyndon Green Junior Mixed School
Lakey Lane Primary School
Northfield Manor Primary School
Kings Heath Junior School
Barford Junior School
Tindal Primary School
Colmore Infants School
Wheelers Lane Infants School
Kings Heath Infants School
Perry Beaches Infants School
Delhurst Infants School

BRISTOL

Trial classes
Glenform Primary School
Holymead Junior School
Parson Street Primary School
Perry Court Junior School
Four Acres Junior School
Chester Park Infants School
Four Acres Infants School
St. Anne's Infants School
Highridge Infants School
Willow Green Infants School

Control classes
Holymead Infants School
Oldbury Court Infants School
Broomhill Infants School
Henbury Court Infants School

CARDIFF—ALL TRIAL CLASSES

Rhiwbina Junior School
Allensbank Junior School

Hywel Dda Junior School
Cwrt-yr-Ala Junior School
Lansdowne Junior School
Llanishen Fach Junior School
Rhiwbina Infants School
Cefn Onn County Primary School
Stacey Road Infants School
Peter Lea Infants School

CARLISLE—ALL TRIAL CLASSES

Denton Holme Junior School
Belle Vue Primary School
Newton Primary School
Inglewood Junior School
Petteril Bank Primary School
Kingmoor Primary School
Belah County Primary School
St. Margaret Mary Primary School
Inglewood Infants School
Newlaithes Infants School

ESSEX—ALL TRIAL CLASSES

Dr Walker's C.E. Primary School
Shelley County Primary School
High Ongar County Primary School
Moreton C.E. Primary School
St. Andrew's C.E. Primary School
Chipping Ongar Junior School
Chipping Ongar County Infants School

ILEA—ALL TRIAL CLASSES

Sandhurst Junior Mixed School
Kingswood Infants School
Beaver's Holt Primary School

KENT

Trial classes
Brook County Primary School
Elham C.E. Primary School
New Romney C.E. Primary School
Willesborough Junior School
Christchurch C.E. Primary School

Hythe Infants School
St. Peter's C.E. Primary School
Sir John Moore Infants School

Control classes
St. Peter's C.E. Primary School
St. Eanswythe's C.E. Primary School
George Spurgen County Primary School

LEICESTER—ALL TRIAL CLASSES

Avenue Junior School
Braunstone Hall Junior School
Eyres Mansell Junior School
Montrose Junior School
Northfield House Junior School
Knighton Fields Junior School
Whitehall Infants School
Overdale Infants School
Willowbrook Infants School
New Parks House Infant School

LIVERPOOL—ALL TRIAL CLASSES

Tiber Street County Primary School
Much Woolton C.E. Primary School
Morrison County Primary School
Childwall C.E. Primary School
Booker Avenue County Primary School
Kingsthorne County Primary School
Dovedale County Primary School
Stocktonwood Infants School
Woolton County Primary School
Craighurst Infants School

ST. HELENS—ALL TRIAL CLASSES

Allanson Street County Primary School
Derbyshire Hill Infants School

WEST RIDING—ALL TRIAL CLASSES

Sutton-in-Craven County Primary School
Earby County Junior Mixed School
Glusburn County Primary School
Skipton Water Street County Primary School
Gargrave County Infants School
Addingham High County Primary School

Scotland

Trial classes
Our Lady of Lourdes Roman Catholic Primary Schoo
Glenlee Primary School

Also taking part
Armadle Primary School
Torpichen Primary School
Blackburn Primary School
Kirkliston Primary School
Kinneil Primary School
Seafield Primary School
Boghall Primary School
Kirkhill Primary School
Linlithgow Public School
South Queensferry Primary School

Information about classes taking part in the trials of the third set of units

Areas	Stage 1 & 2 units Pattern 1	Pattern 2	Stage 3 units
England and Wales			
Anglesey	4		1
Birmingham		16	20
Bradford	7		2
Bristol		16	
Cardiff		16	
Carlisle	6		
Croydon		16	
Essex	6		
Gloucestershire		14	1
Kent		18	
Leicester		16	14
Liverpool		18	21
ILEA		22	8
Somerset		10	
St. Helens	6		3
Staffordshire		17	
Teesside		22	
West Riding		16	
Southampton			18
Scotland			
Dundee	5		
Lanarkshire	5		
Roxburghshire		4	
West Lothian	6	4	
TOTALS	45	225	88

General information about classes taking part in trials of Stage 1 and 2 units of the third set

Type of school	Infants only	11%
	Junior mixed and infants	86%
	Junior boys only	1%
	Secondary Comprehensive	1%
	Middle	1%
Size of school	8 classes or less	41%
	More than 8 classes	59%
Catchment area largely	Urban	47%
	Suburban	43%
	Rural	10%
Children's back-ground generally	Prosperous	15%
	Average	65%
	Disadvantaged	20%
Age of school buildings	Less than 10 years	23%
	10 to 50 years	33%
	More than 50 years	44%
Number of children in class	35 or more	37%
	Fewer than 35	62%
	Cooperative teaching situation	1%
Age range of children	Infant	10%
	1st year junior	6%
	2nd year junior	16%
	3rd year junior	20%
	4th year junior	42%
	11–12 years	6%
Sex of teacher	Male	55%
	Female	45%

List of schools involved in trials of the third set of units

England and Wales

ANGLESEY—STAGE 1 AND 2 UNITS

Caerleiliog County Primary School
Newborough County Primary School
Ty Mawr County Primary School
Rev. Thomas Voluntary Primary School

STAGE 3 UNITS

Sir Thomas Jones County Secondary School, Amlwch

BIRMINGHAM—STAGE 1 AND 2 UNITS

Greenholm Junior and Infants School
Northfield Manor Junior and Infants School
St. Gerards R.C. Junior and Infants School
Clifton Road Junior School
Nansen Junior and Infants School
Tindal Junior and Infants School
Quinton C.E. Junior and Infants School
Perry Beaches Junior School
King's Norton Primary School
Broadmeadow Junior School
Greston Junior School
Colmore Road Junior School
Four Dwellings Junior School
Green Meadow Junior School
St. Benedicts Junior School

STAGE 3 UNITS

Woodcock Hill Junior School
Greet Junior and Infants School
King's Heath Junior School
Hodge Hill Comprehensive School
Brandwood Secondary School, King's Heath
Slade Primary School, Erdington
Wattville Junior School
Tile Cross Junior & Infants School

BRADFORD—STAGE 1 AND 2 UNITS

Great Horton Junior High School
Hutton Middle School

Delf Hill Middle School
St. John the Evangelist R.C. School
Clayton C.E. Junior School

STAGE 3 UNITS

Tong Comprehensive School
Great Horton Junior High School

BRISTOL—STAGE 1 AND 2 UNITS

Holymead Junior Mixed School
St. Gabriel's C.E. Junior Mixed School
Teyfant Junior Mixed School
Westbury Park Junior Mixed and Infants School
Bank Leaze Junior Mixed School
Waycroft Junior Mixed School
Two Mile Hill Junior School
Begbrook Junior School
Sefton Park Junior Mixed School
Wansdyke Primary School
Chester Park Junior School
Our Lady of the Rosary R.C. Junior Mixed School
Air Balloon Hill Junior Mixed School
West Town Lane Junior School
Ashley Down Junior Mixed School

CARDIFF—STAGE 1 AND 2 UNITS

Allensbank Junior School
Cwrt-yr-Ala Junior School
Lansdowne Junior School
Llanishen Fach Junior School
Rhiwbina Junior School
Glanyrafon Junior School
Gladstone Junior School
Kitchener Junior School
Penybryn Junior School
Hywel Dda Junior School
Bryn Hafod Junior School

CARLISLE—STAGE 1 AND 2 UNITS

St. Margaret Mary Primary School
Morton Park Junior School
Upperby Junior School
Inglewood County Junior School

Robert Ferguson Junior School
Belah Primary School

CROYDON—STAGE 1 AND 2 UNITS
Woodside Junior Mixed School
All Saint's Junior Mixed School
Roke Primary School

ESSEX—STAGE 1 AND 2 UNITS
Walton County Primary School
St. John's Green County Primary School
St. Mary's C.E. Primary School
North County Primary School
St. Thomas Moore's R.C. Primary School
The Mayflower County Primary School

GLOUCESTERSHIRE—STAGE 1 AND 2 UNITS
The Leaze Junior School
Joys Green County Primary School
Westbury-on-Severn C.E. Primary School
Plump Hill County Primary School
Lydney County Junior School
Manorbrook County Primary School

STAGE 3 UNITS
Highworth Secondary School

ILEA—STAGE 1 AND 2 UNITS
St. Joan of Arc Primary School
Charles Lamb Infants School
Charles Lamb Junior School
Highbury Quadrant Junior Mixed School
Canonbury Junior School
Rotherfield Junior Mixed School
Canonbury Infants School

STAGE 3 UNITS
Brecknock Primary School
Beavers Holt Primary School
Walworth School
Tulse Hill Boys Comprehensive School
Sandhurst Junior Mixed School

KENT—STAGE 1 AND 2 UNITS

Hilltop County Primary School
Chattendon County Primary School
St. Peters and St. Margaret's Junior School
Luton County Primary School
Hartley County Primary School
Raynehurst County Primary School
Dover Road County Junior School
Benenden C.E. Primary School
Bishops Down County Primary School
Napier County Primary School
Wakeley County Primary School
Byron County Junior School
Barnsole County Primary School

LEICESTER—STAGE 1 AND 2 UNITS

Uplands Infant School
Willowbank Infant School
Forest Lodge Infant School
Eyres Monsell Junior School
Marriott Junior School
Whitehall Junior School
Catherine Junior School
Northfield House Junior School
Knighton Fields Junior School
Overdale Junior School
Coleman Junior School
Forest Lodge Junior School
Hazel Junior School
Avenue Junior School
Shaftesbury Junior School

STAGE 3 UNITS

Alderman Newton's Boys School
Beaumont Leys Secondary School
City of Leicester Boys School
Corpus Christi Preparatory School
Ellis Secondary School
English Martyrs School
Linwood Boys Secondary School
Mary Linwood Girls Secondary School
Mundella Boys Secondary School

New Parks Boys Secondary School
Rushey Mead Boys Secondary School
Lancaster Boys Secondary School
Hamilton Secondary School

LIVERPOOL—STAGE 1 AND 2 UNITS

Barlows Lane County Primary School
Broadgreen Junior Boys School
Lister Drive County Primary School
Corinthian Avenue County Primary School
Clint Junior Boys School
Morrison County Primary School
Mosspits Lane County Primary School
Northway County Primary School
Monksdown Road County Primary School
St. Cleopas County Primary School

STAGE 3 UNITS

New Heys Comprehensive
St. Ambrose Barlow Secondary Modern School

ST. HELENS—STAGE 1 AND 2 UNITS

Windlchurst County Primary School
Robins Lane Infants School
Sherdley County Primary School
Sutton C.E. Junior School
Derbyshire Hill County Primary School

STAGE 3 UNITS

Parr County Secondary School
St. Cuthbert's R.C. Secondary School

SOUTHAMPTON—STAGE 3 UNITS

Central Middle School
St. Mary's Primary School
Swaythling Primary School
Mansel Middle School
Tanner Brook Middle School
Newlands Middle School
Londshill Primary
Holy Family Primary
Wimpson Middle School
Foundry Lane Middle School
Shirley Middle School

St. Mark's C.E. School
St. John's Primary School
Bassett Green Middle School
Bevvis Town Primary School
Mansbridge Primary School
Bitterne Manor Primary School
Redbridge Primary School
Aldermoor Middle School
Moorhill Primary School

STAFFORDSHIRE—STAGE 1 AND 2 UNITS
Manor County Primary School
Coton Green Primary School
Holmcroft Junior School
Nursery Fields County Primary School
Charnwood Junior School
Marshbrook Infants School
St. Nicholas C.E. Primary School
Blakeley Heath County Junior School
Birches Infants School
Endon Hall Primary School
Leek County Primary School
Westwood Road Primary School
Blackshow Moor C.E. Primary School
Streetly County Primary School
Doe Bank Junior School
Hundred Acre Wood Junior School

TEESSIDE—STAGE 1 AND 2 UNITS
St. Joseph R.C. Infant School
West Dyke Infant School
Lakes Infant School
Newcomen Infant School
Thorntree Junior School
St. Mary's R.C. Junior School
Newcomen Junior School
West Dyke Junior School
Lakes Junior School
Newport Junior School
Linthorpe Junior School
Green Lane Junior School
Kader Junior School
Whinney Banks Junior School

Archibald Junior School
Newport Infant School
Archibald Infant School
Clara Lady Dorman Junior School
Clara Lady Dorman Infant School
Green Lane Infant School
Kader Infant School
Whinney Banks Infant School

WEST RIDING—STAGE 1 AND 2 UNITS

Glusburn County Primary School
Aire View County Primary School
Steeton County Primary School
Sutton County Primary School
Elslack County Primary School
Barnoldswick County Primary School
Earby County Primary School
Christ C.E. Primary School
Parish C.E. Primary School
Ings County Primary School
Gargrave County Primary School
St. Stephen's R.C. Primary School
Bradley County Primary School

Scotland

STAGE 1 AND 2 UNITS

Ladywell Primary School
Kirkshaws Primary School
Townhill Primary School
Canberra Primary School
Berryhill Primary School
Balbardie Primary School
Boghall Primary School
Windyknowe Primary School
Torpichen Primary School
Deanburn Primary School
Blackness Primary School
Trinity Primary School
Edenside Primary School
Easton Primary School
Rosebank Primary School
West March Primary School
Kinill Primary School

Information about classes taking part in trials of the unit *Children and Plastics* **Stages 1 & 2 and Background Information**

Area	Number of classes
Birmingham	15
Bristol	12
Cardiff	13
TOTAL	40

GENERAL INFORMATION ABOUT THE TRIAL CLASSES

Type of school	Infants only	11
	Junior mixed and infants	10
	Junior mixed	19
Size of school	More than 8 classes	31
	8 classes or less	9
Catchment area of school largely*	Urban	17
	Suburban	27
	Rural	0
Children's background generally*	Prosperous	8
	Average	31
	Disadvantaged	8
Age of school building (1 unknown)	Less than 10 years	5
	10 to 50 years	20
	More than 50 years	14
Age range of children	Infant	11
	1st year junior	8
	2nd year junior	5
	3rd year junior	5
	4th year junior	15
Grouping of children in age range of trial class	Streamed	3
	Partially streamed	3
	Unstreamed	34

* 2 were ticked when there was a mixture.

CLASSES TAKING PART

Birmingham
The Abbey R.C. Junior Infant School
St. Vincent's R.C. Junior Infant School
Strechford Junior Infant School
Lakey Lane Junior Infant School
Green Meadow Junior School
St. Laurence Infant School
Colmore Junior School
Broadmeadow Junior School
Perry Beeches Infant School
Greenholm Junior Infant School
Delhurst Infant School
Rookery Junior School
Percy Shurmer Junior School
Chandos Junior Infant School

Bristol
South Street Infant School
Bridge Farm Junior Mixed and Infant School
Whitehouse Junior Mixed School
Easton Road Junior Mixed and Infant School
Wansdyke Junior Mixed and Infant School
Air Balloon Hill Junior Mixed School
Ashley Down Junior Mixed School
Holymead Infant School
West Town Lane Junior School

Cardiff
Allensbank Junior School
Pen-y-bryn Junior School
Lansdowne Junior School
Cwrt-yr-Ala Junior School
Glan-yr-Afon Junior School
Hywell Dda Junior School
Rhiwbina Infant School
Gladstone County Primary School
Peter Lea County Primary School

Classes taking part in trials of *Like and Unlike* (Stages 1, 2 & 3)

Teesside
Kader Infant School
Kader Junior School
Green Lane Infant School
Green Lane Junior School
Whinney Banks Infant School
West Dyke Infant School
Dormanstown Infant School
Dormanstown Junior School
Dormanstown Secondary School
Newport Infant School
Nepwort Junior School
Archibald Infant School
Archibald Junior School
Linthorpe Junior School
Redbrook Primary School
Newcomen Junior School
Newcomen Infant School
St. Joseph's Infant School
St. Mary's Junior School
Lakes Infant School
Lakes Junior School

West Riding
Skipton Ings County Primary School
Skipton Parish Church School
Sutton County Primary School
Silsden Aire View County Primary School
Barnoldswick C.E. School
Barnoldswick Gisburn Road Junior and Infant School
Glusburn County Primary School
Gargrave Infant School
Earby Junior School
Bradleys Both County Primary School

Leicester
Mellor Junior School
Northfield House Junior School
Mellor Infant School
Humberstone Infant School
Uplands Infant School
Corpus Christi Secondary Modern School
Lancaster Boys Secondary Modern School

Classes taking part in trials of *Science, Models and Toys* **(Stage 3)**

Liverpool
New Hays Comprehensive School
Ellergreen Comprehensive School
St. Ambrose Barlow Secondary Modern School
Fazakerley Secondary Modern School
Queen of All Saints Secondary Modern School
St. Anne's Secondary Modern School (Girls)
Arundel Comprehensive School
Our Lady of the Assumption Secondary Modern School
Speke Comprehensive School
Wellington Secondary Modern School
Glan Alyn Secondary Modern School
Liverpool Institute High School for Girls
Margaret Beaven Day Special School

St. Helens
Rivington Comprehensive School
St. Anselm's R.C. Comprehensive School
Parr Mount C.E. Comprehensive School

Kent
St. George's Middle School, Sheerness
Danley Middle School, Sheppey
Lady Anne Cheyne Middle School for Girls, Sheerness
Sir Thomas Cheyne Middle School for Boys, Sheerness

Classes taking part in trials of *Ourselves* **(Stages 1 and 2)**

Bristol
Headley Park Primary School
Parson Street Primary School
South Street Primary School
Stockwood Green Primary School
Two Mile Hill Primary School
Westbury Park Primary School
St. Gabriel's C.E. Primary School

Index